The Physics of Birds and Birding

The Physics of Birds and Birding

The sounds, colors and movements of birds, and our tools for watching them

MICHAEL HURBEN

PELAGIC PUBLISHING

First published in 2025 by
Pelagic Publishing
20–22 Wenlock Road
London N1 7GU

www.pelagicpublishing.com

The Physics of Birds and Birding

Copyright © 2025 Michael Hurben

The moral rights of the author have been asserted by him
in accordance with the Copyright, Designs and Patents Act 1988.

All rights reserved. Apart from short excerpts for use in research or for
reviews, no part of this document may be printed or reproduced, stored in
a retrieval system, or transmitted in any form or by any means, electronic,
mechanical, photocopying, recording, now known or hereafter invented or
otherwise without prior permission from the publisher.

https://doi.org/10.53061/VRKD6914

A CIP record for this book is available from the British Library

ISBN 978-1-78427-307-1 Pbk
ISBN 978-1-78427-308-8 ePub
ISBN 978-1-78427-309-5 PDF

EU Authorised Representative:
Easy Access System Europe – Mustamäe tee 50, 10621 Tallinn, Estonia,
gpsr.requests@easproject.com

Cover image: *Navigating Birds* © Kyle Bean, first published in *Scientific American*

Typeset in Caslon by BBR Design, Sheffield

5 4 3 2 1

To Claire, Alex and Anthony

Contents

Acknowledgments ix

Preface: Looking Up 1
1 At the Feeder: Birds, Mathematics and Symmetry 10
2 In the Garden: Hummingbirds, Flowers and Forces 24
3 On the Open Seas: Length Scales, Migration and Molecules 37
4 In a City Park: Movement, Murmuration and Magnetism 50
5 By a Forest Pond: Impacts, Waves and Sounds 66
6 Under Night's Cover: Hearing, Recording and Analyzing Birdsong 81
7 At the Lake: Sunlight, Reflection and Refraction 98
8 During a Big Day: Light, Matter and Feather Colors 112
9 In the Blind: Images, Eyes and Cameras 131
10 From a Great Distance: Lenses, Binoculars and Scopes 149
11 Under Extremes: Heat, Cold and Thermoregulation 169
12 Above the Earth: Wings, Lift and Flight 182
Coda: Looking Back 198

Notes 201
Glossary 221
Bibliography 234
Index 238

Acknowledgments

This book exists only because of the influence and help of others. Although I developed an affinity for science and nature as a child, I also dropped out of high school. Luckily, my friend Eric Maranowski persuaded me to attempt college despite this, and it worked out. It was my privilege to learn physics at the University of Utah from the very best, including Emeritus Professor Bill Sutherland, winner of the 2019 Heineman Prize for Mathematical Physics. Professors Robert Kadesch, Owen Johnson and Orest Symko were equally instrumental in my development. At Colorado State University, Professors Richard Eykholt and Carl Patton, my Ph.D. advisor, pushed me to a significantly higher level.

I wish to thank Anne Janzer and Dr. Daniel Tortora for their encouragement and advice when this project was in its infancy. As the book matured, Skye Loyd provided valuable commentary as well. I am indebted to Dr. Angela Hornsby and Brie Ilarde at the University of Minnesota for their assistance with the specimen collection at the Bell Museum of Natural History. Editors Lisa Glass and Rob Colson offered invaluable feedback.

Emeritus Professor James Van Remsen of Louisiana State University is one of the great ornithologists of our time, and I am honored to have received his insights and comments on portions of the text. Professor Paul Schaeffer of Miami University (Ohio) also provided helpful discussions relating to the biological aspects of this work. My friend Dr. Richard Cox helped me to restrain my inclination to digress, and you can thank him for persuading me not to include a section on the Higgs boson.

'For me, the study of these laws is inseparable from a love of Nature in all its manifestations. The beauty of the basic laws of natural science, as revealed in the study of particles and of the cosmos, is allied to the litheness of a merganser diving in a pure Swedish lake, or the grace of a dolphin leaving shining trails at night in the Gulf of California...'
—Murray Gell-Mann (© The Nobel Foundation 1969)

'We should always endeavor to wonder at the permanent thing, not at the mere exception. We should be startled by the sun, and not by the eclipse. We should wonder less at the earthquake, and wonder more at the earth.'
—G.K. Chesterton

Preface: Looking Up

First optics

In 1979, when I was 12 years old, I made the most important purchase of my life. I handed my paltry savings, plus a contribution from my parents, to the man behind the glass display case of the camera bar in our local K-mart. In exchange, I clutched my very first optics. I was too young to realize that what I could afford was inevitably of miserable quality. The long-defunct brand, Focal—made exclusively for K-mart—sounded as top-shelf to me then as Swarovski does now.

As soon as I directed my gaze—through this not-quite-precision instrument—out at the natural world, I was enthralled. Sights previously beyond view were now revealed in detail and color. So many vivid marvels were waiting to be seen, appreciated, checked off a list. How magical it was that distant things could be brought within reach, via nothing more sophisticated than several carefully shaped pieces of glass, judiciously arranged.

I needed a book to help me identify what I was seeing; I chose a *Peterson Field Guide*. Soon, every relevant title from the library had cycled through my hands. This led to another revelation: no less rewarding than seeing a new and magnified world was the pleasure of learning about it from books. The nomenclature, the jargon, the binomial scientific names aroused not anxiety, but fascination. There was a beautiful logic to the systematic structure.

I began to draw what I was seeing, which led to a curiosity about photography. I somehow acquired a used 35 mm Single-lens Reflex (SLR) camera, and learned about f-stops, exposure times, the art of capturing light and how to develop film. Our bathroom became a part-time darkroom. The pungent odors from the trio of chemicals—developer, stop-bath and fixer—were not popular, but I loved them. They gave the house an erudite, adventurous character.

I obsessed over the weather and the tricky business of forecasting it, for time outdoors was dictated by the mood of the atmosphere. I came to disapprove of rain, snow and the clouds that could thwart my photographic intentions.

Springtime was the most anticipated season in my adopted hometown of Salt Lake City, not only for the welcome thaw but also for the return of the stunning variety that went missing in the winter. With luck and effort, one might even check off over a hundred targets during a single 24-hour marathon event.

I frequently asked to go further afield, away from the city, as there was so much more to see. But my parents were not amenable to driving out of town every weekend and waiting in the car as I deployed my optics and my camera. They did, however, let me join a local club, and soon I had a network of people willing to give me a lift to outings in the mountains or deserts of Utah. I envied those adults, not just for their better optics, but because they had the means to travel. Some spoke of trips to South America, where they saw things that were impossible to glimpse at home.

Big nights

You might think I've been recounting an early discovery of birding but, please forgive me, because I've been stringing you along. In 1979, I was pointing my first telescope at other colorful denizens of the skies, and my first paperback of the Peterson series wasn't the monumental handbook so instrumental in making birding a popular and accessible hobby; it was an early edition of *A Field Guide to the Stars and Planets*.

That my story could have described either avian or celestial obsessions illustrates just how much the two pursuits have in common. Like birding, amateur astronomy flourished during the 20th century, with the advent of relatively inexpensive optics. Long before the term 'citizen science' was coined and sites such as eBird and xeno-canto existed, nighttime naturalists had blurred the line between hobby and meticulous observation: they were finding new comets, discovering supernovae and making myriad contributions to our shared knowledge. What follows is a run-down of all the similarities I've alluded to in the previous section.

Telescopes and binoculars are central to the two passions: matters such as field of view, eye relief and focal length concern both camps. Photography, with its endless combinations of components and settings, can be an obsession whether your quarry is stars or birds. And whereas clouds can lead to less-than-optimal, back-lit bird photos, they simply make astronomy impossible. We can bird in the rain and snow, but without clear skies, an astronomical telescope is just glass and a tube, revealing nothing.

Astronomy naturally produces communities of enthusiasts, much like local branches of the National Audubon Society. Groups typically host 'star parties', where members gather before sunset, set up their scopes and share the night sky. My hometown club, the Salt Lake Astronomical Society, would usually hold them beyond the city so as to escape light pollution, the glare from countless streetlights and billboards that blots out all but the brightest celestial objects.

World travel beckons the astronomy buff because the stars seen from deep in the Southern Hemisphere are forever hidden from those living in the far north (and vice versa). From the vantage point of most people, the earth always obstructs the view of the Southern Cross, Alpha Centauri (a moniker that illustrates how stars, like species, also have binomial scientific names) and the Magellanic Clouds. Only when approaching the Equator will they rise into view as our pole star and two 'dippers' are lost under the northern horizon.

Seasons play a role too, as the parade of constellations slowly advances throughout the year, making the skies of different months unique. By chance, springtime brings the Northern Hemisphere astronomer an uptick in celestial variety; that is when most 'deep-sky objects'—the galaxies, star-clusters and nebulae—happen to be visible. For many, these are the most sought-after targets in the skies—the wood-warblers of the celestial world, holding an undeniable eminence and challenge.

For a few weeks in late March and early April, it is possible to observe all 110 members of a celebrated collection of deep-sky objects known as the Messier (MEZ-zee-eh) objects, over the course of a single night. The name honors the astronomer who made the first comprehensive catalog of these objects in the late 18th century. In the analog to a Big Day of birding, the dedicated will work their telescopes from sunset to sunrise, attempting to see each of them, an event known as a Messier Marathon. How many astronomers realize that such spring nights are full of living, multicolored jewels coming from the south, invisibly passing between them and the stars that steer their migration? And how many birders take stock of the immense celestial variety that lies so far beyond the birds on clear spring nights, waiting to be revealed through their optics?

Physics, birds and vision

As a child, I directed telescopes and binoculars only up at the night sky, never at the terrestrial world. The biological world didn't attract my interest, somehow. Many sky-gazers looking to expand their science horizons will turn to physics, even if they do not pursue it formally or have mathematical inclinations; it did, after all, evolve in concert with astronomy. The observed motions of the moon and the planets, studied by Galileo, Kepler and later, Newton, helped establish the field of mechanics, the earliest branch of physics. The co-location of astronomers with the physics faculty persists in many universities.

Physics came to supplant astronomy for me. It didn't just explain the clockwork night sky, the shapes of galaxies and the colors of the stars that first lit my interest in science; it explained (or attempted to explain) everything. It underpinned chemistry and, by extension, geology, meteorology, biology and ecology. It was the one field to unite them all, with a deep abstract beauty manifested in every aspect of reality. This is a view shared by almost all physicists. Ask us for a one-word description of what we are so enamored with, and most will respond 'Nature'. What we mean is, 'Nature at its deepest level, that which underlies everything.'

On a spring afternoon, during my doctorate program, I took my work to a city park, hoping for inspiration for some problem I was stuck on. But my focus on calculations was broken by the antics of a woodpecker on the ground, and I needed to know what this creature was called. I stopped my work, went to the bookstore and picked up my second Peterson guide. The bird was called *Colaptes auratus*, or Northern Flicker, and it would be the instigator of a kind of madness. As I was doing physics when it started, I still cannot decouple physics and birding.

The happy discovery of birds was followed by one far less pleasant. As I was wrapping up graduate school, I began losing my eyesight. Night vision was the first to go, and this was rudely demonstrated during a star party when I couldn't see the faint wisps of nebulae that others were raving about. My field of vision was becoming more constricted as well, forming a kind of tunnel. I'd soon be diagnosed with retinitis pigmentosa, a genetic condition that destroys the function of the rod cells that enable peripheral and low-light vision.

I'm fortunate. I still have a smidgen of vision left, a field of view roughly the size of a fist held out at arm's length. The last scraps of light are precious to me. Decades have passed since I've seen the faint light of a Messier object in a telescope, but with effort and patience and luck, I can yet get cooperative birds into my narrow range of vision. I work hard to see them because I have no choice; my obsession has not subsided in the face of disability.

I'm also fortunate to have had gifted professors who taught me to see the world in a way that doesn't require eyesight: from the physics perspective. This book is my humble attempt to share the view from this prospect, to illuminate the nature behind birds in a different way.

Hitched to everything else

Birders are curious people, in both senses of the word. We are statistical outliers, set apart by an obsession with feathered dinosaurs that is not widely shared or understood. And we are inquisitive; not merely about what might be skulking around the next bend, but also in seeking explanations for the striking sights, sounds and behaviors of birds. Why are breeding male warblers so brilliantly colored? How do they learn their songs? What causes them to migrate such enormous distances? We ask the questions instinctively and find their clear and substantive answers deeply pleasing.

It is impossible to pursue birds, or answers to our questions about them, without grasping the larger ecological systems that they inhabit. The panorama of wildlife, beyond the avian, patiently waits for and rewards any attention we give it. We needn't take birds off their apex position to expand our scope, and the happy work of identification applies no less to alders and willows than to the flycatchers perched on them.

To be a birder is to be a naturalist. What a lovely, expansive moniker that is. It is a big tent we share with the botanist engrossed with orchids, the lepidopterist with their fritillaries and skippers; and, yes, even the rock-hound who celebrates the agate. The inanimate world, the necessary bedrock of the living one, is of no less consequence. A procession of towering clouds crossing the plains, a sandstone canyon opening out to distant mesas, a black and bright night star-scape: these readily awaken a sense of awe. But beyond such demonstrations of primitive beauty, many face a barrier. The world beneath all of this—the physics that makes it possible—may seem too abstract, esoteric, perhaps even a bit boring. But it isn't.

Some treatments of 'hard' sciences such as physics and chemistry might leave us cold, but the fault isn't with the subject matter. We can partly blame evolution, which endowed us with the senses and discernment that are requisite to survive, but no more. To grasp the depths of nature we have to get past a frontier imposed by our coarse native senses, and develop intuition for non-intuitive things. It takes work.

The naturalist with little background in the physical sciences is liable to be frustrated in several ways. A dense, inexplicable tangle of equations might well deter them from ever wondering what secrets they hold. But having to settle for only facile answers in lieu of deeper explanation is equally unrewarding. For example, to be told that the reds and blacks of feathers arise from different pigments, and to leave it at that, is deeply unsatisfying. We are left to wonder, 'how do pigments function, exactly?' And 'specialized glands' might be the 'answer' to the question of how seabirds survive without fresh water, but this superficial response tells us nothing about the fascinating chemistry going on up there over those tubular noses.

'When we try to pick out anything by itself,' said environmentalist John Muir, 'we find it hitched to everything else in the universe.' The harrier is hitched not just to the hilly field it patrols, but the gravitational field it exploits. The underlying minutia, the fundamental processes that power all things, should also interest us, for they are no less beautiful than the birds. To illustrate this, and to make the surprised naturalist glad of this fact, is a goal of this book.

Euclid alone?

'Euclid alone has looked on Beauty bare,' wrote Edna St. Vincent Millay, and her poem is without equal as a glowing hymn to the austere delicacy of geometry. But the declaration in that first line couldn't be more in error. Euclid may have seen it first, but he is far from alone. The allure is there for everyone to see, equally, should they look for it.

Grasping a scientific principle—figuring out some feature of the world that had been beyond comprehension, or working out a mathematical proof for oneself—forges a profound, personal connection to nature. And it doesn't matter that you were not the first to do it. When you prove a geometry theorem or derive a key physics result, and you truly understand what you've done, it then 'belongs' to you no less than to the individual credited with its discovery. It's the same as how a quality encounter of a Wilson's Warbler *Cardellina pusilla* in the field is no less momentous now than it was for Alexander Wilson, or anyone else before or since.

There is a dual kind of gratitude that this should inspire. Gratitude that we can, with modest effort, come to stand beside the brilliant and experience an understanding no less sublime, and gratitude for the existence and efforts of the geniuses who made the trails accessible for the rest of us. Great scientists, no less than great artists, deserve their acclaim, for they are endowed with uncommon gifts of insight. The artist builds something that would otherwise not exist, but it requires no less creativity for the scientist to unravel nature's riddles. The end result might 'only' reveal what was out there all along, but it takes as much inspiration to show others what nature so artfully conceals.

Consider yourself invited to collect insights into the inner workings of nature in the same way you might collect bird sightings, pressed flowers or dazzling agates. Write them down, track them and review them. 'The scientist does not study nature because it is useful; he studies it because he delights in it,' said Jules Henri Poincaré, 'and he delights in it because it is beautiful.' One need not be a scientist to share in this delight.

Galileo's heirs

There is another source of tension, for some of us at least, between the concerns of naturalism and the steady progress of the physical sciences in the form of technology. And this ties back to the limits of our innate perceptual abilities.

Fortunate as we are that natural selection gave us sensory tools, it only endowed us with just enough to permit our ancestors to survive. Our unaided eyes are quite weak. Not only are there countless stars and galaxies that we cannot directly perceive, the microscopic constituents that compose us are also beyond direct experience. We are oblivious to ranges of signal beyond the red and the violet, and sense but a tiny fraction of the vast spectrum of radiant energy cascading down all around us, as if great booming music were being played on a vast pipe organ, but we only perceive sound when just a few of the many keys are pressed.

Only when we started seeing beyond our intrinsic limits, with the development of optical instruments in the early 1600s, could science begin in earnest. The development of the telescope and the microscope, at roughly the same time, extended our perception in dramatic ways. Every technology that expands horizons and enables the scientific odyssey—from parabolic dishes to antennas to cosmic ray detectors—is a descendant of a pair of lenses held in a tube.

Our indispensable appliances of birding are not very different from that first telescope wielded by Galileo. Every time we lift our binoculars and magnify the image of a bird, we are re-enacting his historic step, extending our eyesight beyond the limitations that evolution imposed, ironically via the intelligence it also granted us.

That the deep unity of nature can only be revealed via technology is something that has never sat well with some naturalists. Henry David Thoreau wrote, 'All our inventions are but improved means to an unimproved end.' If this was an understandable reaction in his newly industrial era, it is more so today amid the generation of plastic trash, heat-trapping exhaust and a digital world enabling a stupefying level of narcissism. But his brush was too broad. If we imagine a time-traveler headed back to Walden, bringing images from the Hubble telescope and some quality binoculars to take into the woods, it is likely that old Henry might revisit his sentiment. Squaring up nature and technology is a theme that will appear again in the following chapters.

About this book

The beneficial accord of technology and birding became most apparent to me during a Neotropical trip long ago, when I first experienced a guide skillfully brandishing a green laser pointer to help his guests locate the gnatwrens and foliage-gleaners darting through dense tangles. It was a revelation, especially as it significantly improved my odds of getting birds into my restricted visual window. What I mostly recall was that I kept laughing; not because something was *funny*, but out of the sheer joy at the simple yet wonderful ingenuity of it.

Some time later, a guide steered my attention up a steep embankment and silently circled a small clump of flowers with the shimmering green dot. I searched for the skulking bird there, but nothing appeared. When I eventually gave up, I asked him what I'd missed. 'Still there!' he said, 'Orchids!' I had no idea why this was supposed to be important, and later, out of earshot, I asked my wife. She rectified my ignorance about these special and uniquely gorgeous plants, and now I ask guides to point them out as well. Such flowers have become, for me, synonymous with any and all non-avian loveliness that pops up during an outing.

In this book, we'll stop to look at the orchids as well.

Imagine we are planning a trip to a vast natural reserve, rich with birds. Some are already familiar to us; others are targets we hope to add to our life list. We understand that some of our quarry will be abundant and easily identified, while other birds will be recalcitrant and challenging. This protected place holds a variety of habitats and is so large that a dozen trail-heads dot the periphery. One path was once a popular roadway, wide and flat for miles. It's now the kind of trail you don't notice, where one can step back for a better view without fear of tumbling into a ditch. Another trail is a narrow staircase of rock and tree root, cutting into the hillside: exactly the kind of path one does notice. Some tracks are muddy and call for Wellington boots; some are boardwalks. Others are scarcely discernible as trails at all.

All these paths form an irregular grid of intersections, sometimes sprouting loops and cul-de-sacs. The connections are so numerous and convoluted that there is no way to survey this place in a linear fashion. We cannot go from point A to point Z and be done with it. Some trails lead to the same locale, but offer different views along the way. We might walk the periphery of the pond shore on one outing, but on another see it from the cliffs above.

The process of exploring such a birding mecca is a metaphor for how this book will go. In exploring the physics of birds and birding, we have much and varied ground to cover. Our targets are not birds, but explanations. Some will be relatively easy, and the trails needed to reach them will be short. Others will require some exertion, and we'll have to go more slowly and have some patience. Sometimes we'll arrive at a place we've been before, but via a different route, and certain features will become familiar landmarks. We'll often reach intersecting trails where we'll have to pick one direction, noting that something equally intriguing was lying to the left when we chose to go right.

As naturalists, we must be prepared to wander down the side trails and take in the rocks, trees and skies. And as birders, we know that a challenging target might require that we put our binoculars—and head—down and focus on the path for a few miles first. Please keep that in mind if you find yourself wondering, 'where are the birds?' We might just be looking at an orchid.

Answering birding-specific questions is a big goal here, but it isn't the only one. The molecular and atomic worlds present ground we must cover if we truly want a more intuitive grasp of color and sound. But in and of themselves, these worlds are as beautiful as any other aspect of nature; we just need some time and a shift in perspective to see it. When we do that, we glimpse one of the most sublime facts about nature: *what we perceive depends strongly on things we are not, and cannot be, directly aware of*.

If I do my job as a guide, I'll help you find some targets, but I'll also illuminate some not-so-obvious connections within the natural world. This preface, I hope, will entice some birders to take their optics out on a clear night and have a look upward. They may find another passion to complement their daytime explorations. Perhaps in the springtime, when the avian and astronomical varieties peak, some eagle-eyed young naturalist will start a new kind of marathon—a back-to-back Big Day of birds and Big Night of deep-sky objects—checking off 110 of each.

Chapter 1

At the Feeder: Birds, Mathematics and Symmetry

We glance out of a window, a flutter of wings catches our eye, and we stop to look at the birds on the feeder. Unconsciously, we begin to identify them, making lists and counting entries. In this chapter, we'll consider how elementary mathematics underpins the natural world, through species counts and yearly cycles, the shapes of waves and eggs, fern leaves and falcon flights, and symmetries at the heart of physics.

Identifying and counting

'To call things by their proper name,' said Confucius, 'is the beginning of wisdom.' It is also a rewarding, addictive pursuit and the very heart of birding. Every tremor in foliage, nondescript chirp or distant speck wheeling against the clouds presages an identification to be performed, a little mystery to be solved.

Birding is rich with the joy of finding things out, naming, categorizing and ticking boxes. Identification requires discrimination, the recognition of differences in sizes, shapes, colors, sounds and behaviors. We may do it without conscious effort, with commonplace birds, or it might involve meticulous study when dealing with difficult or unfamiliar ones. It is a matter of discerning differences that often have a quantitative character. To assess the length of a bill in proportion to the head is to make a rough measurement. To judge a feature as 'more rounded' is to engage in a geometric study. To estimate the relative wingbeat frequencies among a mixture of waterfowl is to appraise and compare the rates of periodic motions. To bird is to measure.

While we differentiate among features so as to identify and name species, we also engage in the opposite process: integration. Without hesitation, we mentally group birds by type and then enumerate them. This may be focused work, such as a meticulous count of a lake's various waterfowl; or an estimate from memory—made hours after an outing—of how many sparrows were in a clearing. And of course, many of us become obsessed with making accurate summations of the species counted over the course of a day or a lifetime. Numbers are fundamental to our hobby.

The ability to count is not endemic to humans. Other animals—insects, fish, amphibians, mammals and of course birds—are capable of discerning quantities.[1] This isn't surprising, as there is an obvious benefit to keeping tabs on the number of predators about, but some skills seem to go beyond this. Irene Pepperberg's famous studies of African Gray Parrots *Psittacus erithacus* demonstrated that they can be trained to acquire numerical skills comparable to those of very young children; and, like some corvids, they can grasp of concept of zero.[2] Other abilities appear hard-wired. Specific neurons in the brains of untrained Carrion Crows *Corvus corone*, for example, have been mapped to the recognition of different numbers.[3] Newly hatched Domestic Chickens *Gallus domesticus* can track counts of objects, apparently via both addition and subtraction.[4]

Perhaps the fact that 'lower' animals can perform addition makes it seem pedestrian, but the fact is, counting by one in an iterative manner is *the* fundamental operation upon which all arithmetic rests. Everything else that we do with numbers is a variation on that theme. Subtraction is merely the reverse of addition. Multiplication is a clever trick to speed up the work of doing sums: 4 sets of 10 sparrows get us to 40 faster than plain counting. And division simply inverts this device, providing a comparison of numbers: as when we note that, say, a quarter of the birds at the feeder were finches.

Beyond these rudiments, of course, extends a vast, sophisticated realm of exponentials, logarithms, trigonometric functions and far more. But no matter how forbidding any of it may seem, or how esoteric advanced concepts can become, they are ultimately reducible to acts of marking off one, two, three, etc. 'There is no problem in the whole of mathematics,' said the Austrian physicist Ernst Mach, 'which cannot be solved by direct counting.'

Simple enumeration may be all we need during a nature outing, but if our goal is to go beyond observation—if we want to understand the hows and whys of birds and birding—then we will be forced to reckon with some higher mathematics. This may appear a daunting barrier for some. Dense formulae are understandably off-putting to those without the specialized education needed to parse them. But just as one can enjoy a sonata without knowing how to read sheet music or play the piano, so too can one appreciate *what the equations are saying* without having to invest years of study in learning *how to solve them*. This is the spirit in which mathematics is presented in this book.

The inseparability of mathematics from the natural world first becomes apparent when we consider its geometrical manifestation. There on the prairie horizon lies the same straight line of Euclid's text; look elsewhere and find a perfect circle in the expanding ripple left by a trout, the six-fold symmetry of a snowflake, the spiral of a nautilus shell. When we peer deeper, at the motions of planets or the bending of light by a lens, the numerical character becomes impossible to ignore.

Galileo referred to mathematics as the 'language' that we must learn in order to read the book of the universe. This is a celebrated but unfortunate metaphor because the similarities are superficial. Myriad languages have been created, consisting of ever-changing rules authored by humans. But symbols are very different things from what they refer to. We invented the letters c, r, o and w, but not the black corvids they are assembled to represent. And mathematics isn't about the marks we make on paper, but the innate relationships among what the marks represent. The patterns of ink are of our own devising, but once we assign them particular meanings, the results take on a life of their own, enacting complex dramas we could never have dreamed up. Mathematics is *discovered*, not invented.

Physics is concerned with the fundamental rules governing matter and its interactions, as revealed by experiment. We inquire, and we find that the rules use numbers. Moreover, the rules are followed to such a staggering degree of accuracy that some are surprised at how 'unreasonably effective' the mathematics is.[5] Should we be astonished? It seems inevitable if we recognize that math is not a mere 'tool' but rather the ultimate substrate upon which everything is built. As cosmologist Max Tegmark put it, the universe itself is a mathematical structure.[6]

Sine waves and egg shapes

What is astonishing isn't that nature employs mathematics, but how economically it does so. Certain patterns emerge in seemingly disconnected physics, pointing to a profound underlying simplicity. Consider the simple example of counting birds. Many of us have tracked species counts in our yard, favorite locale or county over time. We might summarize our results in a table, but such a display isn't edifying. A more insightful way to explore our records is to produce a graph, allowing the numbers to manifest in a geometrical way, making them easier to grasp. An example of five years of species counts from a single North American county is shown in Figure 1.1.

Two features jump out from this graph: regularity and variability. An underlying repetition is ornamented with a smaller level of changing detail.

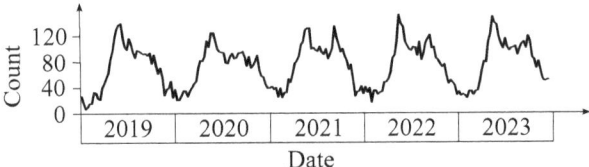

Figure 1.1 Species counts for Sherburne County, Minnesota, over the course of five years, based on four tallies per month. [Source: eBird.⁷]

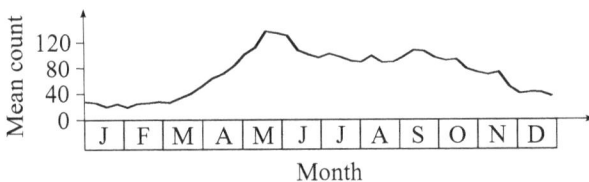

Figure 1.2 Average (mean) species count over a yearly cycle, with four measurements per month. Based on the five years of data shown in Figure 1.1.

We understand the latter because, for any given week, there are confounding factors that influence what we see or hear; that is, the weather, changes in populations, level of effort and, of course, luck. But the background rhythm dominates this picture, and we're drawn to seek out the underlying theme. We might average the numbers to reveal how a typical year 'should' look.

Figure 1.2 shows the average (mean) count, with its smoother trend. To begin with, only the hardiest birds that are adapted to winter are present. Species variety grows rapidly in the spring, then rebounds back down, as some birds continue north, while others remain to nest. Another uptick occurs in the fall, as the northern breeders pass through again; but with less fanfare of song and plumage (and more leaf cover), we tend to record fewer species than in the spring.

That we should take an average is so self-evident that we don't think to justify it. It works because, while any particular date this year is a different point in time from the same date last year, it is yet identical in a relative way: on these two dates, the orientation of the earth relative to the sun is the same.

This orientation determines the intensity and duration of daylight, and hence the total energy available for plant growth and reproduction, as well as temperature, which drives the level of biological activity. Migration is the obvious consequence for animals that are capable of pursuing the regularly shifting abundances of resources that ensue. In temperate zones, the variation in solar energy and its effect on ecology is dramatic. We might capture this by mapping out how the maximum angle between the sun and the southern

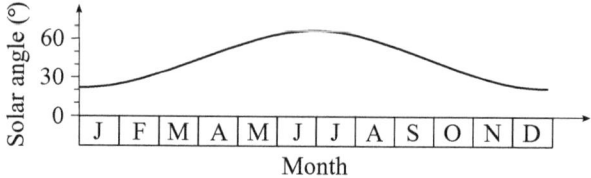

Figure 1.3 Approximate maximum angle of the sun relative to the southern horizon over the course of a year, based on 45° latitude.

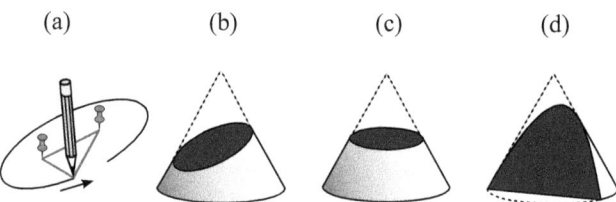

Figure 1.4 (a) Construction of an ellipse, by means of drawing with a pencil constrained by a string that wraps around the pencil and two fixed pins. (b) A cone, cut at a shallow angle, also produces an ellipse in cross-section. (c) A perpendicular cut on the same cone yields a circle. (d) A sufficiently steep cut on the cone produces a parabola.

horizon varies day by day, because the sun's height correlates to the energy delivered. At a latitude of 45°, we would find that the approximate solar angle over the course of a year changes as shown in Figure 1.3.

Although this curve differs from the averaged species count of Figure 1.2, it is undeniably connected, for it captures migration's root cause. It has the approximate form of the ubiquitous *sine wave*, or *sinusoid*, which is found everywhere in nature and which will appear repeatedly throughout this book. (Why it is 'approximate' will become clear shortly.) To explain why *this* shape occurs, we need to consider the details of the earth's orbit.

As the earth completes a circuit around the sun, it traces out the shape of an ellipse: a flat, oval-like loop. Figure 1.4 (a) shows how an ellipse can be drawn using a pencil, two pins and a loop of string that is kept taut around them, forming a triangle. Diagram (b) reveals how the same shape can be found as a *conic section*, in which a cone is sliced at a shallow angle to leave a cross-sectional cut that is elliptical in shape. Make the cut perpendicular to the cone's axis, as shown in (c), and a circle results; a sufficiently oblique slice (d) produces a very different curve, the *parabola*. We remark on these related shapes not only because they all 'reside' in the cone, but because they all describe motion in the presence of gravity. Ellipses and circles are traced out by planetary orbits, while a parabola marks the path of a tossed ball, or a bird with its wings folded in, falling for a spell during its undulating flight.

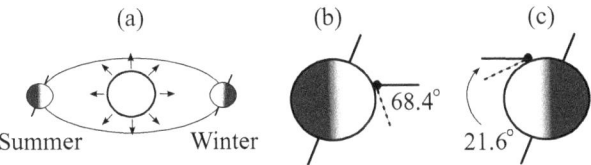

Figure 1.5 (a) Schematic of the earth's orientation relative to the sun, showing its rotational axis tilted relative to the plane of its orbit. The two cases shown correspond to its position at the summer and winter solstices. (b) The position of an observer at 45° north latitude is indicated by the dot, and the large solar angle between the southern horizon (dashed line), and the midday sun (solid line) is evident, during summer. (c) The same construction for winter, where the angle is greatly reduced.

The ellipse and circle show an obvious kinship, so it isn't surprising that they have numerical connections. For example, the area of the ellipse is found by multiplying its height by its width, dividing by four, and then multiplying by the constant π (pi), with its famous value of 3.14159... π is a transcendental number without end, defined as the ratio of a circle's circumference to its diameter, and is deeply intrinsic to nature. If mathematics were a mere human invention, we'd simply *set* this value; perhaps choosing something simpler, such as 3. But we have no say in it. Once we identify what 'circumference' and 'diameter' mean, we are left to discover their resulting ratio. And the fact that π cannot be written out in full is even more startling. A kind of infinity hides in every circle and ellipse.

For the earth's elliptical orbit, the sun sits at one focus (corresponding to the location of one of the two pushpins in Figure 1.4 (a)), so our distance from the sun varies throughout the year. There is a widespread misunderstanding that this drives seasonality on earth, but it has nothing to do with it. Our orbit has only a slight elliptical shape (its width and height are within a few percent of each other) and we can justifiably approximate it as a circle going forward. It is the inclination of the earth's axis at 23.4° relative to the plane of the orbit that matters. This is because, on one side of the orbit, the Northern Hemisphere tilts toward the sun; and then six months later, it is canted away from it, as shown in Figure 1.5 (a). Diagrams (b) and (c) construct the maximum and minimum solar angles for an observer at 45° north latitude. The dashed lines show the direction of the southern horizon, and the solid lines show the direction of the sunlight.

This allows us to see how the highest and lowest solar positions occur (at the summer and winter solstices, respectively) but how do we get the sine wave of Figure 1.3? To eschew the math, we'll exploit a clever geometrical analogy.[8] To represent the orbit, we take a circle and extend it to make a cylinder; and, to account for the inclination of earth's axis, we tilt it. In Figure 1.6 (a) we put a handle on this canted cylinder and dip it into a level pan of paint. In (b) we use

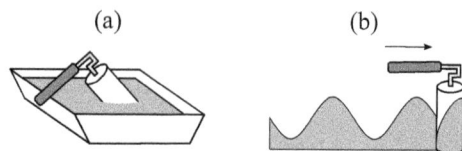

Figure 1.6 (a) A cylinder is tilted and partially immersed in paint. (b) Using it as a roller to apply the paint to a flat surface, a sine wave is produced. [Redrawn from Apostol and Mnatsakanian (2007).]

it as a roller, running it along a flat surface, where it will leave a border in the form of the undulating sinusoid. The sine wave is an inescapable component of circular motion, but to reveal it we must project onto one dimension fewer, as the cylinder does to the wall. Note that we'd need a paint roller with an elliptical cross-section to emulate our actual orbit, and the resulting wave would have a somewhat distorted character. Hence, our measured solar angle is an approximate sinusoid. Had we not tilted the roller upon dipping it, we'd have a straight border—just as a planet with an uncanted axis and a circular orbit would have no seasons.

The sinusoid has properties that are crying out for nomenclature and numbers. The extent to which the wave rises and falls vertically is the *amplitude*. The duration of one cycle of the wave, if measured in time, is the *period*; whereas, if we measure it as a function of space, we refer to it intuitively as *wavelength*. This will be discussed further in later chapters. A sine wave such as in Figure 1.6 (b) is in various ways symmetrical. Draw a line through its middle, and excursions above and below that line will be mirror images. Take a single wavelength and translate it the same distance, and we've changed nothing. Periodic functions like this are amenable to subdivision into finite, identical 'units', one of which tells us everything about the entire chain.

Nature ever breaks into repeating patterns and movements, across any scale we might look at, from the orbits of planets down to the structure of matter. A feather tract on a bird shows periodicity in location, a repeated copy and paste. Each position gets translated from the previous one, with the staggered character of overlapping shingles. We see periodicity in the still photo of a windswept lake, with regular peaks and troughs in endless succession; or in the weave of a fiber in an oriole's nest, ever rising over and falling under crisscrossing strands.

Periodicity can be more abstract. In 1869, not every chemical element had been discovered; but, for those that had, their relative weights and behaviors in chemical reactions were known. The Russian chemist Dmitri Mendeleev set out the elements in order of increasing weight and saw repeatable patterns in their reactivity. From lithium, for example, moving through the next eight

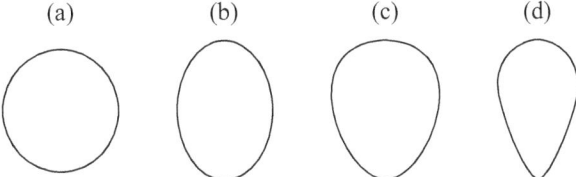

Figure 1.7 Examples of the four egg shapes generated by the newly discovered equation that can produce them all: (a) spherical, (b) ellipsoid, (c) ovoid and (d) pyriform.[9]

elements, we arrive at similarly behaving sodium; eight more steps take us to potassium, which has comparable properties. Mendeleev did not know what caused this regularity, but he had the insight to rearrange the elements into a two-dimensional table, boldly leaving spaces for some that had yet to be discovered (and that later were). His famous periodic table ultimately derives from geometrical considerations of how electrons orbit the atomic nucleus. This is a consequence of the attractive electrical force that leverages the same form as the gravitational law dictating planetary motion. More on this in Chapter 2.

Returning to the ellipse, its shape reminds us of an egg in profile. Indeed, if we could spin our ellipse about its horizontal axis, it would create a three-dimensional ellipsoid, which is the form of the eggs laid by species such as the flightless Maleo *Macrocephalon maleo*.[10] Additional egg shapes can be found in other species, of course, but, in total, eggs have been classified to fall into four different groups: spherical, ellipsoid, ovoid and pyriform.[11]

It may seem surprising, given the simplicity of an ellipse, that a single equation that can describe all four egg shapes was discovered only in 2021.[12] It isn't particularly edifying to look at, so we'll skip the math and show representative examples calculated with it in Figure 1.7. Why different species utilize particular egg shapes continues to be studied. Researchers in 2017 showed a clear connection between wing morphology and egg asymmetry.[13] Birds that have longer, narrower wings tend to produce eggs that are less spherical and more elongated, mirroring their more streamlined body shape. But why various flightless birds produce widely differing egg shapes isn't addressed by this research.

Even though the equation governing the seemingly elementary geometries of eggs is rather complicated, it is generally the case that the smooth shapes in nature have correspondingly simple mathematical underpinnings. And, conversely, an equation describing a complicated relationship—such as in Figures 1.1 and 1.2—would typically require many terms and possess a similarly 'messy' quality. We might guess, then, that only the trivial geometries of the natural world—such as the knife-edge of the ocean's horizon, the solar disk or the curve of a hanging vine—will be expressible with elementary connections between numbers, and that complex shapes will require inelegant rules and copious mathematics. This is a mistaken assumption, as we'll see next.

Ferns and fractals

Figure 1.8 shows a fair representation of a fern leaf, or frond. There are no smooth curves here; rather, the image consists of 30,000 points. We can imagine creating this portrait by hand, referencing a sample picked up from the forest floor, and dabbing the paper with a fine pen. Marking three dots per second, we'd be done in just under three hours.

But that isn't how this likeness was made. It was created on a PC running a few lines of code, based on simple equations discovered by mathematician Michael Barnsley in 1993.[14] To generate such detail from elementary rules seems implausible. The secret lies in the concept of iteration, where an equation takes numbers as input and returns others as output, which are then fed back into it in a repetitive process. Let a computer tally the values from a simple recipe, and the fern shape gradually appears when we use those numbers to plot points on a grid.

A Barnsley fern uses four variations of one rule to make different leaf parts: the 'stem' comes from one version of the equation, while the leaflets arise from the others. The rule specifies how to take any pair of numbers (i.e., a unique point on the grid) and find the next point. Specifically, for the point (x, y), the rule for getting the next point, (x_{new}, y_{new}), is:

$$x_{new} = ax + by + e$$
$$y_{new} = cx + dy + f$$

The coefficients a, b, c, d, e and f are fixed numbers, or constants, which differ for the four variations. The details are not important here, and we only show these equations to emphasize that the math is quite simple: multiply, add, plot the point and repeat.

Figure 1.8 Thirty thousand points approximating a fern leaf, generated with the Barnsley fern algorithm.

Figure 1.9 A sample of the infinite detail of the Mandelbrot set.

If we did not have such a rule to generate our image, we would have to capture or quantify it by making a list of 60,000 numbers, like a bitmap file specifying pixel after pixel. Instead, it is all encapsulated by a rule that is just a few keystrokes long. The detail pops out of a few *linear* equations; we need not speak of squaring, finding square roots, exponents, trigonometric functions or anything like that.

For those previously disinclined to see mathematics lurking in the natural world, come on in, the water's fine.

This leaf is an example of a *fractal*; that is, a type of irregular shape with a self-similar character. Fractals abound in nature: in lightning bolts, cloud borders, the branching of river deltas and circulatory systems, and so on. Yet their mathematics was discovered only recently. The 'mother of all fractals' might be the *Mandelbrot set*, named after mathematician Benoit Mandelbrot, who produced the first detailed illustrations of it around 1980.[15] A portion of this dazzling object is shown in Figure 1.9.

Like the fern leaf, the Mandelbrot set emerges from a simple rule applied at all the coordinate (x, y) points that form the image. The rule is:

$$x_{new} = x^2 + x - y^2$$
$$y_{new} = 2xy + y$$

How does this work, exactly? Suppose we start with the point (0,0): plugging in for x and y, the rules give us (0,0) right back. Continuing won't change the outcome. On the other hand, suppose we start with the point (1,0). The rule will lead us next to (2,0), which upon further iteration gives (6,0), then (42,0), then (1806,0) and so on. We say (1,0) 'runs away' because it grows ever larger, while (0,0) 'stays bounded.' For those starting points that stay bounded, we might mark them white; and those that run away, we mark black. Essentially, that is all there is to it, but to make an image, we must look at many (x, y) pairs,

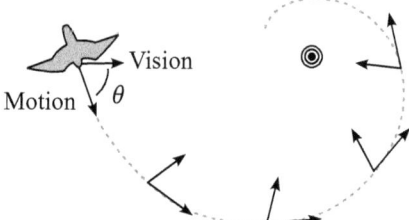

Figure 1.10 A schematic (not to scale) of a falcon tracking prey (indicated with the 'bullseye') located at an angle θ relative to its direction of motion. If the angle is to be maintained, the flight direction must change as the distance is reduced. The result is a logarithmic spiral, from the point of view of the prey.

determine if they run or stay, and plot them accordingly. Doing such work by hand is terribly laborious, but the task is perfect for a computer. Hence the study of fractals got going only as the necessary computing power became available, starting in the 1980s.

Figure 1.9 contains various grays to help bring out the detail. The shade is determined by tracking *how rapidly* each unbounded point grows in size. If this is done in full color—perhaps with blue for fast and red for slow, with an entire spectrum in between—we'll produce an even more striking visualization. A series of images made with increasing zoom, or resolution, can then be strung into videos that portray 'descents' into the fractal. Dramatic examples of this are found on YouTube (search for 'super hard Mandelbrot zoom') and Wikipedia, and the reader is kindly asked to put down the book and pull these animations up. The challenge in watching the otherworldly and unlimited beauty unfold is keeping in mind that *the generating mathematical rule is absolutely trivial*. You will never view math the same.

Note how various patterns in Figure 1.9 are repeated. One obvious feature, right of center, resembles a seahorse tail. Search for more and you'll find they are legion. This shape, ubiquitous in nature, is a *logarithmic spiral*; that is, a mathematical relationship that is not part of the explicit rules generating the Mandelbrot set. It has been shown that Peregrine Falcons *Falco peregrinus* will trace out a logarithmic spiral in pursuit of their prey.[16] Why don't they simply dive along a straight line? The answer draws on several considerations. First, raptors' visual acuity is highest when looking to the side, not straight ahead. This is because of the location of the *fovea*—the portion of the retina with the greatest resolution—which we will discuss in Chapter 9. It behooves the falcon to keep its prey 'in the cross-hairs' of the fovea, so the bird must either turn its head or turn its body as it approaches. We illustrate the latter case in Figure 1.10. Starting at the most distant position, the falcon has the 'bullseye' lined up to its fovea, along a line that makes an angle θ relative to the flight motion. We immediately see that in order to maintain this angle upon approach, the falcon must continually alter its flight direction; and this inexorably traces out the logarithmic spiral pattern, from the perspective of the prey (which is itself moving, of course).

Why the doesn't the falcon simply turn its head to the side during the pursuit, keeping the fovea locked on target while flying straight? Because doing so would present a larger, less sleek obstacle to the onrushing air, increasing the resistive drag force and slowing the bird. We will encounter drag later, in the contexts of diving birds (Chapter 5) and flight (Chapter 12). The upshot is, the higher speed that the longer spiral path affords will put the falcon on its prey sooner than the straight—but slower—flight would. The principle of *least time* is such a fundamental strategy that even inanimate nature employs it; it governs the propagation of light and underpins all our optics, as we will see starting in Chapter 7.

That most of us tend to find fractal designs attractive is confirmed by psychological studies that show a preference for images correlated to their *fractal dimension*: a well-defined number that captures 'how fractal' a given shape is.[17] It is a curious thing. We expect that evolution would shape our aesthetics for biological ends; attraction to the human form, or the pleasing sound of running water are obvious examples. But why the affinity for vistas of distant, jagged mountains or the branching cracks of lightning bolts? It isn't clear, but it is a capacity to be grateful for.

Perhaps it isn't just us; a recent study suggests that the Red-legged Partridge *Alectoris rufa* of Western Europe might recognize and use fractal shapes.[18] It was shown that the level of self-similar detail in the pattern of their bibs, as calculated as a fractal dimension, was correlated to immune responsiveness; reduced food intake and a decrease in well-being also affected this number. If more-detailed bibs indicate healthier individuals, it is not unreasonable that appraisal of this feature could play a role in mate selection.

Symmetry and conservation

We have one final, if more abstract, illustration of how mathematics and geometry inform concepts that we'll see repeatedly. It is an example of the sublime natural architecture that physics reveals and that physicists revel in.

Consider a chemical reaction. We combine several substances and later find that different chemicals have been produced. The amounts of the various chemicals change during the process, but the net amount of material must stay the same. This accords with our expectation that matter won't just appear or disappear. Mass in a chemical reaction is governed by a simple *conservation law*; the total is a fixed, unchanging quantity.

Such laws might seem trivial, but they can be incredibly useful. Consider energy, which, like mass, we have some intuitive feel for. It is formally defined as *the capacity to do work*, and work can be done in various ways, such as the application of a force across some distance. Energy can be stored as chemical potential energy—for example, in gasoline, batteries or sunflower seeds. Alternatively, potential energy can be manifested through position, wherein rocks atop a cliff have greater gravitational potential energy than similar rocks below them. When stored energy is released—say, when the rocks fall—kinetic energy, the energy due to an object's motion, is generated.

Energy can be messy—taking on all these different forms—which is why the principle of energy conservation is welcome. When we account for all the transformations it can make, we find in the end that the ledger is always balanced. The falling rock takes potential energy and changes it into kinetic energy. The conservation rule tells us they must be equal, allowing us to do things such as predict the final speed based on the starting height.

Momentum, which we can express as velocity multiplied by mass, obeys a conservation law. The fact that moving bodies tend to maintain their motion, while stationary bodies remain at rest, demonstrates that momentum has an unchanging character. This rule is indispensable when applied to a group of objects, for the total momentum cannot change, provided that no external force is acting. This allows us to predict the outcomes of collisions, for example.

Closely related, and similarly conserved, is angular momentum. This is a property that is manifested in spinning objects—from ice-skaters to planets to phalaropes—and is determined by mass, rotational speed and shape. The repeatability of the sunrise demonstrates the conserved angular momentum of the earth, as does the fact that stationary bodies do not suddenly start spinning.

That these conservation laws have practical application is certainly important, but that is not why we are on this digression. Physicists are as deeply (if not more) concerned with knowing *why* such laws exist. That they do have an origin was discovered by the scientist whom Albert Einstein called 'the most significant creative mathematical genius' of her time, Emmy Noether.[19]

Noether's story is not widely known. A Bavarian Jew like Einstein, she would also escape National Socialism by coming to the United States in 1933. Unlike her famous counterpart, she endured relentless sexual discrimination throughout her career. As a student, she was forced to ask permission to attend classes; as a faculty member, she was only allowed to teach without pay. She was prolific despite these obstacles and the brevity of her 53 years on earth.

Noether built a vast and important body of work in mathematics, but it was her contribution to modern physics, *Noether's theorem*, that may be the most profound yet simple idea in science that most people have never heard of. It can be stated thus: *every conservation law is a direct result of a fundamental symmetry.*

A definition of symmetry can be hopelessly abstract, more easily visualized than described with words. Per Merriam-Webster, it is 'a correspondence in size, shape, and relative position of parts on opposite sides of a dividing line or median plane or about a center or axis'.[20] Most birds are symmetric in a bilateral way, where the left side mirrors the right. Notable exceptions include the crossbills, where each individual has an asymmetrical bill, although populations on average have as many 'top-left' as 'top-right' birds.[21] The Wrybill *Anarhynchus frontalis* of New Zealand all have both their mandibles turned to the right.[22] Many Australian parrots show asymmetry in behavior, with more individuals favoring their left foot.[23]

Other common symmetries involve rotations. A columbine flower, for instance, has a five-fold symmetry; turn it 72°, and it looks the same. A circle has continuous symmetry: rotate it by any amount and it appears unchanged.

It isn't just a circle that demonstrates perfect rotational symmetry: the entire cosmos does. That is, there appears to be no preferred direction to space as a whole; there is no cosmic 'This Side Up' sign. Were the entire universe to be (gently) rotated by a quarter turn to the left, say, there would be no way of knowing that such a singular event had occurred. Put another way, the experimental results obtained in a laboratory would not change if we were to rotate all the equipment by the same amount.

Noether's theorem tells us that this rotational symmetry must lead to the conservation of angular momentum. The earth continues to spin because there is no preferred direction in space to 'break' the symmetry. When we consider translational symmetry—the fact that were the entire universe slid (again, gently!) to the right by a few meters, nothing would change—Noether comes back to us with momentum conservation. A boulder stays put, and an asteroid hurtling through space continues unabated, because there is no 'preferred' place for them. And finally, from the time translation symmetry of the universe—the fact that physical laws are the same today as they were yesterday—her theorem proves that energy must be conserved as a result. Stated differently, one need not decree that the universe has both symmetries *and* conservation laws. The latter are logically inescapable given the former.

There are other, subtler symmetries that are foundational to physics, but we'll close with these.[24] Noether demonstrates—as do the Barnsley fern, the falcon's spiral and other examples we'll see throughout this book—that mathematics deeply permeates everything around us, and recognizing this will only enrich our experience of birds and nature.[25]

Chapter 2

In the Garden: Hummingbirds, Flowers and Forces

Next, we visit a sunlit garden where hummingbirds make their signature mockery of gravity. Less obvious are electrostatic forces enabling the transfer of pollen, creating mutual benefits for plant and bird. In this chapter, we'll consider these interactions and how they connect with the magnetism that birds exploit for navigation. And we'll see how a celebrated flower led to arguably the most important scientific discovery ever made.

Force and gravity

If there is one avian family capable of provoking an interest in physics, even in the most science-averse, it would be Trochilidae—the hummingbirds. Their dazzling, iridescent colors, frenetic metabolisms and unique manner of feeding from flowering plants all plead for explanation. But nothing is more striking than their signature ability to remain suspended in midair. There is something uncanny about a stationary body sustained on whirring wings that beat too rapidly to be seen.

Were a Calliope Hummingbird *Selasphorus calliope* to alight on one tray of a balance, we could bring it to level by a setting a penny on the other.[1] Bird and coin each feel the same attraction to the earth—the downward force of their *weight*—because both possess the same amount of *mass*. Mass is what provides *inertia*—that is, a resistance to change in motion—and it quantifies the amount of 'stuff' making up an object.

Inertia is indispensable to the laws of motion, which are fundamentally about mass, distance and time. Multiply a body's mass by its velocity to obtain momentum, which we introduced in Chapter 1, in the context of a conservation law. When no external force is acting on an object, its momentum cannot change—that is, if it is at rest, it will stay put, and if moving at constant speed, it will maintain that motion. Noether's theorem tells us that this is because any place in space is 'as good' as any other, that there is a translational symmetry to the world. This is the famous first law of motion deduced by Isaac Newton.

Only when some force is applied to a body can the momentum change: a child kicks a stationary ball, and it moves. The *rate* at which the momentum changes is, conveniently enough, equal to the force. When the mass of an object remains the same during such a process, as is usually the case, the only part of momentum that can change is velocity; this means that *acceleration* occurs. Hence, the familiar second law from Newton, simply stated as: *force is mass times acceleration*. These relationships can be concisely expressed as:

$$\boldsymbol{F} = \frac{d\boldsymbol{p}}{dt} = m\boldsymbol{a}$$

The force $\boldsymbol{F}$, momentum $\boldsymbol{p}$ and acceleration $\boldsymbol{a}$ are in bold type to indicate that they are *vector* quantities; that is, they have direction as well as magnitude. The mass, m, does not have any kind of directional bearing; it is just a number, or a *scalar*. As for the term d/dt, it indicates the *rate of change* of the variables that follow it. If the force is zero, then the rate of change of momentum is zero. The momentum need not be zero but, whatever it is, it will stay unchanged, or conserved.

Momentum conservation applies not only to single objects, but also to collections of them. If a loaded rifle is at rest, its total momentum must remain zero, even after it is fired. The bullet gains tremendous momentum and so the rifle must carry an equal amount, which manifests as recoil in the opposite direction. The same holds for the rates of change: the rate at which the bullet gains momentum must equal the rate at which the rifle's momentum increases. Hence, the forces must balance as an action–reaction pair: equal and opposite. This is Newton's third law.

We experience the third law in hammering a nail: the nail moves into the wood, the hammer bounces back up. One impact causes two transitory, contrary kinds of motion. It should be intuitively clear that something similar must occur with each forward and reverse stroke of a hummingbird's wings. Although the physics of flight is not a trivial topic—we will devote the final chapter of this book to it—we can make some simplifications here in order to get a preliminary sense for it.

A very basic model for the motion is shown in Figure 2.1. Diagram (a) shows the figure-eight pattern made by each wingtip. During the forward stroke (b), the wing pushes a parcel of air down and to the left. In (c), the

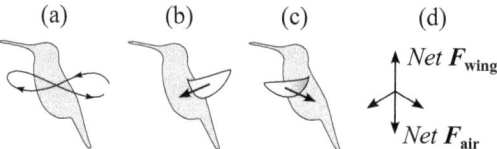

Figure 2.1 (a) The figure-eight pattern made by the wingtip during a complete cycle. (b) The forward stroke forces a parcel of air down and to the left, shown by the arrow. (c) The reverse stroke pushes air down and to the right. (d) The net force on the air is the sum of the forces from the two strokes, which leaves no net horizontal component. The wing experiences a net upward reaction force.

reverse stroke pushes air down and to the right. The forces on the air from both motions are added in (d), with the horizontal components canceling each other out. The air is driven down, and the wing experiences a net upward force.

The lift obtained from a single stroke of the wing, like the recoil of a hammer, is transient. This is simply the practical restriction of having a finite wing or arm that can only move so far. When the limb can move no further, there is no way to generate any more force. Meanwhile, the ever-present pull of gravity is relentless, so the bird must continue working in order to counter its weight. Without the continual effort, it would accelerate downward at about 10 meters per second per second, just like any other falling body.

This business of generating force by pushing off, the concepts of action and reaction, and the inertia that keeps bodies at rest or moving along, do not strain our intuition. But it took revolutionary genius to first grasp these rules. The three laws of motion alone are sufficient to warrant Newton's almost divine status as the father of physics.[2] But an arguably greater accomplishment was his understanding of gravity. His insight—that the same mechanism causing an apple to fall must also hold the moon in orbit around the earth (and the earth around the sun)—was nothing short of astounding. It may seem obvious now, but until 1687, nobody had put it together.

Newton's dissection of gravity laid bare the minutiae of how it depends on mass and distance, as captured by the law of gravitation. For two masses, m_1 and m_2, separated by a distance r, the force lies along a straight line between their centers and is proportional to both masses, divided by their separation squared. This is all said more elegantly as a mathematical expression:

$$F = -G \frac{m_1 m_2}{r^2} u$$

The term u is a 'unit vector' and simply indicates that the force is directed along a line connecting the two masses (or their centers). G is a necessary constant that accounts for our choice of measurement units, such as kilograms and meters, and the minus sign indicates that the force is always attractive. Combine this with the previous expression for force—as mass times acceleration—and we'll obtain an *equation of motion*, whose solutions are the circles, ellipses and parabolas from Chapter 1.

Gravity is a *fundamental force*; that is, it is an interaction intrinsic to nature, a built-in feature that cannot be changed. There are four such forces (that we know of), and the existence of each is integral to How Everything Works. We'll look at another fundamental force momentarily, and the other two in later chapters.

Although Newton deduced the mathematics of gravity, he couldn't fathom *how* it could possibly work. The earth and the moon (or the apple) are not in contact—how could a force exist between them? This is called 'action at a distance' and it can be disquieting to ruminate upon—what, exactly, is transmitting this force? Newton admitted that the idea of gravity acting between distant bodies, separated by space with nothing to mediate it, was 'an absurdity'. He would not publicly advance any ideas about how it might operate, and his contemporaries ridiculed the idea.[3] An explanation would not come until the mid-19th century, when the physics of electricity—no less relevant to our hummingbird than gravity—was being worked out.

Charge and pollination

Hummingbirds are practically defined by their unique relationship with plants. They must hover in order to feed, because delicate flower stems wouldn't support a perched bird, or because their tube-like structures would be otherwise inaccessible. The nourishment is offered up in exchange for the birds' role in the plant's reproductive strategy, shuttling pollen from flower to flower. In the process, they leverage another kind of action-at-a-distance force, one that isn't as readily observed as gravitation.

A flower extends its pollen out for an insect or avian courier by bearing it upon anthers at the ends of long filaments. But the anthers needn't rely on physical contact with the visitor for the transfer to occur. The grains can leap a modest gap between flower and pollinator, and readily cling fast to the new host, because they are charged and experience an electrostatic force.[4]

That some materials can accumulate electrical charge due to contact with other objects has been known since antiquity. Thales of Miletus described how fossilized tree resin attracted other items after being rubbed with animal fur.[5] The Greek word for this material, which we call amber, is *elektron*. But only in the mid-1800s did anyone quantify such effects and scrutinize the interactions that would come to plague clothes tumbling in a dryer. Charge was found to come in positive and negative varieties, as described by Benjamin Franklin, the celebrated statesman, scientist and denigrator of Bald Eagles *Haliaeetus leucocephalus*, which he claimed suffered from 'bad moral character'.[6] This bipolar nature leads to several far-reaching consequences.

In 1785, Charles-Augustin de Coulomb worked out the details of how charges interact, discovering the law that bears his name:

$$\boldsymbol{F} = k\frac{q_1 q_2}{r^2}\boldsymbol{u}$$

This should look familiar; it is identical in form to Newton's gravitation law. As with gravity, force decreases rapidly as distance increases. The charges q_1 and q_2 stand in for the masses, and a different proportionality constant, k, replaces G. That such different phenomena should utilize the same mathematics is an example of the great economy of nature and is something that physicists get excited about. But something novel results from the fact that charge can be positive or negative, whereas mass is always greater than zero. If the charges have the same sign, the force is positive: that is to say, repulsive. Otherwise, the force will have a negative value: demonstrating, as they say, that opposites attract.

Charge may seem foreign to us, because we usually interact with matter that is apparently uncharged. But this is because of an additive, canceling effect. There is a tremendous amount of charge around us and in us: about as much positive as negative, effectively mixed up together, so that the total charge (and hence force) becomes zero. Electrical effects typically become apparent when accumulations of the two varieties of charge are spatially separated.

Although the forces of gravity and electrostatics obey laws having the same form, their relative strengths are wildly different. We might be tempted to think that the ubiquitous gravity—rudely keeping us pinned to the earth—is the larger force, but the reverse is true. Consider how we stand motionless upon the ground, as every last gram of material in the earth aims to pull us down via gravity. We are prevented from sinking only because the ground beneath us pushes back up. It does so by virtue of the electrical repulsion between 'our' electrons (bound to the atoms of the soles of our feet) and those comprising the matter that supports us. But this patch of ground—only a tiny fraction of all the earth's material—generates a comparable force via electrical charge. It must be intrinsically much stronger than gravity. To put a number on it, we can simply consider two protons, which have mass and are therefore attracted via gravity, but are also repulsed by their identical charge. Calculating the two forces using the laws above, we find that the electrical repulsion is 10^{36} times stronger than the gravitational attraction. Written out in full, that's 1,000,000,000,000,000,000,000,000,000,000,000,000.

Despite this potency, electrical effects often cloak themselves via the intermixing of positive and negative, and typically become apparent only when we separate the charges. The timeworn recipe from Thales is a fine way to divide and collect charge: merely place certain items (such as cat fur and glass, or hair and a latex balloon) in contact and rub them. This process, known as *tribocharging*, causes an accumulation of negative charge on one object and positive on the other, because different materials possess different affinities for positive and negative charge. These differences exist because some atomic structures welcome additional electrons, while others seek to lose them.

The earth and its atmosphere undergo a constant cycle of tribocharging. Rising warm air masses and falling precipitation provide a frictional effect similar to walking across a thick carpet. Positive charge accumulates in the atmosphere, negative charge on the earth's surface. The process occurs gradually and out of sight, except for the occasional violent rearrangement of charge known as an electrical storm.

The excess negative charge on the earth distributes itself across its surface and onto whatever resides there, including plants, flowers and their pollen grains. Fortuitously, the materials making up the bodies of birds and insects favor a net positive charge. By virtue of their motion in the air and tribocharging with other parts of the environment, they can arrive among the blooms primed to make electrostatics play an ecological role.

The effect of charge on pollination has been studied extensively with insects.[7] Through measurement of the charge they can carry, and models of the forces acting on pollen grains—electrostatic, gravitational and air resistance—a compelling picture emerges: pollen transfer isn't just a matter of the grains being 'sticky' (although some varieties have burr-like projections that help keep them mechanically attached)—electrostatic attraction can be key to the adhesion process.

As a pollinator approaches the anthers, the separation decreases and the force acting on the charged grains increases rapidly, per Coulomb's law. The pollen may experience enough attractive force to leap to the pollinator without physical contact. Once aboard, the pollen will become more positive, soaking up the ambient charge of its host. This is no less important, for there is a second transaction to be completed. The pollen now has the inclination to return to the negative environment of a subsequent flower, hopefully finding its way to the stigma, where it can finish the fertilization process.

It shouldn't be surprising, then, that this mechanism can play a role with hummingbirds as well as insects. Researchers studying Anna's Hummingbirds *Calypte anna* have shown that they can carry more charge than insects typically do.[8] Under typical conditions, there's enough 'static cling' present that pollen grains can jump gaps of several millimeters between flower and bird. It is a serendipitous trick—a hidden cycle of flowing charge, exploited by plants to cycle their genes, ensuring their next generation will bloom and feed the hummingbirds—ultimately powered by the convection of air masses driven by solar energy.

Charges exert forces on one another and move about, so we can speak of their energies as well. A charge will have a certain potential energy depending on its position among other charges. For example, when a negatively charged pollen grain is approached by a positively charged hummingbird, it has a high potential energy, similar to that of a rock teetering on cliff's edge. This potential for charge is what we mean when we refer to *voltage*. The term might seem endemic to electrical engineering and its inventions, but it shows up everywhere, including in cell biology.

Charges at high voltage possess more potential energy than those at lower voltage. And just like a mass raised up high, they will return to 'ground' if not impeded from getting there. The charges in the power lines that stretch across the landscape experience high voltages, but they are blocked from flowing to ground because they have no path to get there; the surrounding air is an insulator, a sufficiently impenetrable barrier.

A bird perched upon a high voltage line experiences no ill effects because it provides no path for the charge to flow down to a lower potential. If the bird were large enough to contact both the line and ground (or a second cable carrying a different voltage), then it would have a problem. Charge would flow through it like water released from a reservoir running down a steep river channel. Biological systems, built from delicate cells sensitive to small voltage differences, are easily damaged when charges go careening through them.

The same term is used to denote the flowing movement of both water and charge: *current*. Mathematically, current is another kind of *rate of change*, measured by counting how much stuff—water molecules or charged particles—goes by per second. But, unlike the flow of water, electrical current automatically creates another kind of force, one cleverly exploited by both human and avian navigators.

Electricity and magnetism

The uncanny push and pull that we experience when handling several magnets provide another palpable demonstration of 'action at a distance'. This fundamental force has been put to much practical use, such as when it guides a sliver of iron set free to rotate and align with the north–south direction of the earth. We've used compasses to navigate since they were invented in China over

1,000 years ago, a small fraction of the time that various birds have utilized their own internal ones. (The avian biological compass, which has only recently been uncovered, is far more complex than a suspended iron needle. It will be covered in Chapter 4.)

As with electrostatics, magnetic forces have a dual character. Instead of positive and negative, the names that capture magnetic polarity reflect the historical importance of its use as a tool for orientation: every magnet has a north and a south pole.

Electricity and magnetism are intimately connected, but this was not realized, surprisingly, until the early 1800s. The link is readily demonstrated by placing a compass near a wire. When electrical current is passed through the wire, the needle will obediently turn in response. This is a very useful trick that leads to many practical applications, such as the production of sound via a loudspeaker. Here, the wire is wound into a coil, which is attached to a lightweight cone that can move. On the other side is a magnet, attached to the speaker enclosure. When a time-varying electrical current (the signal) is passed into the coil, it will act like a magnet whose properties change in lockstep. There will be a resulting push and pull between the coil and fixed magnet that moves the cone, thereby pumping out sound waves that replicate the electrical signal in the wire.

Does this mean that a magnetized bar of iron is somehow animated by electrical currents? No, because magnetism also arises from atomic properties of certain materials. We typically refer to a bar of iron as a *permanent magnet*, as opposed to a coil of wire carrying a current, which is an *electromagnet*. They produce identical effects, but the latter can be easily turned off or varied in strength.

The earth itself, of course, is a huge magnet. Whence comes this geomagnetism? Not from vast stores of iron, but rather from currents. Below the crust and silicate mantle, a central volume approximately the size of Mars consists of liquid metals around a solid core. The great heat of this—which is continually fed by radioactive decay—drives convective motions, not unlike the rise and fall of water coming to a boil. The earth's turning imparts a rotational impetus to the charge-carrying fluid, causing cylindrical flows.[9] It's as if a huge coil were connected to a battery.

That is the picture in broad strokes; the details are far more involved. The motion is difficult to model (and impossible to observe directly) but is expected to result in anomalies and changes in geomagnetism. The north and south poles have flipped many times over geological time spans, and changes within smaller time-frames have also been seen. A recent study shows how it can affect bird migration.[10]

Eurasian Reed Warblers *Acrocephalus scirpaceus* winter south of the Sahara and make long treks to Europe to breed. In 2019, researchers studied almost 18,000 banding results dating from 1940 to 2018, and noted that individual birds didn't always return to the same locations. What they found was that breeding grounds shifted in concert with changes to the local magnetic *inclination*: the degree to which a compass needle deviates from the horizontal. The migrating birds were using weak magnetic signals to determine where to halt their journeys. We'll unravel how they are thought to do this in Chapter 4.

After magnetism was found to arise from charges moving through a wire, something like the reverse was discovered by the British physicist Michael Faraday. Moving a magnet near a loop of wire, he observed the process of *induction*, whereby a change in the magnetic environment causes an electric current to flow. This effect enables all manner of technology from generators to microphones. But Faraday was far more than the grandfather of numerous practical inventions. He was a genius and visionary of Newton's caliber. This is all the more impressive given that he was almost entirely self-educated and unskilled at mathematics. He had tremendous intuition and insight that made up for this, and his skills as an experimenter were unmatched. Perhaps it was his unconventional background that led him to his most profound contribution: he was the first to explain how 'action at a distance' worked, by showing that it is mediated by a *field*.

The definition of a field is necessarily abstract: it is a continuous mathematical entity, taking on different numerical values across its extent throughout space and time. Examples are helpful, and a simple one is temperature. At any place on earth, a thermometer will read some value—a measure of how hot or cold the local air is. Walk sufficiently far and you'll see the reading go up or down. Wait a while and the value will eventually change, because most fields—certainly the more interesting ones—change over time. While we cannot practically determine the temperature at every possible location on earth, it is clear that there *is* a temperature everywhere, measured or not.

Another meteorological example of a field is wind velocity. Unlike temperature, which is given by a single number (it is a *scalar field*), the wind requires a number to denote its speed, and at least one more to represent its direction (for example, '6 mph, from 30° west of north'). The wind is a *vector field*, often visualized by a map with arrows at various points.

For these examples of meteorological fields, we must, of course, have a planet with an atmosphere. But there are other fields that are *fundamental*; they permeate all of space and exist as a brute fact. It was two of these, the *electric* and *magnetic fields*, that Faraday initially postulated. And, by extension, the long-range interactions that exasperated Newton were then understood to be mediated by a *gravitational field* that filled all of space, connecting planets, moons and everything else.

The interdependence of electrical and magnetic effects indicated that they were tied at the hip. The two fields would soon be found to be part of a single entity, the *electromagnetic* (EM) field. If this unity wasn't pleasing enough, an even bigger surprise was lurking. When James Clerk Maxwell pulled together the four equations that describe the EM field, he found that they predicted that *waves* could also ripple and spread through it, at a breakneck speed of roughly 300,000 kilometers per second: the speed of light. Such waves, in fact, *are* light. Everything we see (or utilize as radio waves, microwaves, X-rays, and more) comes to us as undulations in a single field that mediates every electrical or magnetic phenomenon, everywhere. The economy of nature is extreme.

Of course, if these EM waves didn't somehow interact with (or couple to) matter, we'd never notice them. Charge is what ties matter and light together. The presence of charge changes the field, and the field in turn creates forces acting on charges. Consider the analogy of a stationary fishing bobber, floating on a still pond, which reshapes the water surface under it. When a fish expertly strips the bait from the hook below, the tugging accelerates the bobber down (before the buoyancy of the water accelerates it back up), which causes waves to spread outward across the surface. Another object floating on the surface—a leaf, an insect, or another bobber—will rock up and down in concert once the spreading waves reach it. So it is that a charge, made to oscillate at some frequency, will create EM waves radiating out at the same rate, with the effect that charges beyond it will be driven to move in accord.

Vision, optics and the physics of coloration all depend on this connection between matter and light, and we will encounter these ideas again and again. We are seeing how the linkages between seemingly disparate things run very deep, and following them has rapidly led us far from our starting point in the hummingbird garden. We'll return to the flowers now for another surprising insight they've given us.

Flowers and atoms

In the Pacific Northwest there grows a type of evening-primrose *Clarkia pulchella* that has collected a variety of charming monikers, including deerhorn clarkia, beautiful clarkia, ragged robin, elkhorn clarkia and pink fairies (see Figure 2.2). It was first described in 1806 during the famous Corps of Discovery Expedition, led by Meriwether Lewis (also known for his woodpecker) and William Clark (also known for his grebe and nutcracker). Several decades

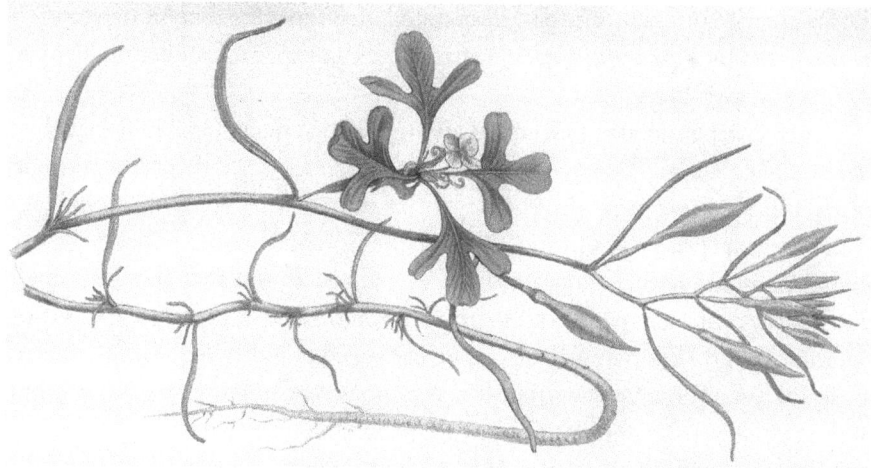

Figure 2.2 An illustration of the *Clarkia pulchella* plant and flower, made by Frederick Pursh in 1813.

later, another naturalist, David Douglas (for whom the fir tree is named), sent some *C. pulchella* seeds to London, writing, 'I hope it may grow in England.'[11]

By 1827, the seeds had produced flowering plants in that far-away place, under the care of Robert Brown, the botanist that first described the cellular nucleus. Eventually, *C. pulchella* pollen grains (about 0.1 millimeters in extent, and visible to the unaided eye) found their way under his microscope. Brown was interested in the mechanics of fertilization, and these samples intrigued him because he could discern, within the grains, smaller components which might play a role.

Immersed in water on a microscope slide, the pollen grains would burst open, spilling their interiors into the surrounding liquid. Their contents included two types of tiny organelles, roughly a hundred times smaller than the pollen: starch-containing *amyloplasts* and lipid-holding *spherosomes*.

Brown called the organelles 'molecules'; but his term meant something different from what it means today. It denoted elementary, organic building blocks theorized to compose all living beings. As he examined them, he noticed that they were constantly moving, first in one direction, then in another, each uncorrelated to the jiggling of the others. He broadened the scope of his study, including pollen from other plants, and found the same behavior. He wondered if he was observing some animate, living force.

His next investigations would be decisive. He fabricated comparably sized bits of inorganic material to study; making powders from glass, rocks and even 'a fragment of the Sphinx'. If the movement was a result of a life process, then these objects—his controls—would show no locomotion. But the same performance on the microscopic stage continued; it was not specific to the organic world.

Brown didn't know what was causing the motion, and eventually gave up and even downplayed the work, considering it of scant value. But the behavior would yet become known as 'Brownian motion'. It would be explained some 80 years later by Albert Einstein, and it would precipitate a dramatically new understanding of reality.[12]

Einstein intuited that the organelles moved in response to random impacts from particles too small to be seen. By making quantitative predictions, his work led to further experiments that produced definitive evidence that the individual components of the water itself were repeatedly striking the organic structures. The importance of this idea cannot be overstated. Einstein's insight had demonstrated the existence of chemical molecules: bonded groups of two or more atoms, the smallest possible amount of a substance.

The great American physicist Richard Feynman made the following analogy for Brownian motion.[13] Imagine a large playing surface, crowded with people. They have a gigantic beach ball, much larger than any of them, and every person is trying to push it in whatever direction. We view the scene from afar, and cannot see any of the individuals, but we see the ball and notice a jerky, random movement. There is no organized principle behind how the people push, but sometimes a majority pushes in roughly the same direction, and we'll see the ball lurch that way, until something else happens.

The jostling water molecules can move the organelles, but not the larger, massive pollen grains. The upshot is that anything that can be seen without optics is too large to be noticeably disturbed. Hence, this clue about the structure of matter came only after the microscope was invented. Brown, like Galileo, used optics to kick-start a revolution that he'd know nothing about.

The explanation for Brownian motion took almost a century to arrive. The idea that matter was built from discrete chunks had long been a conjecture, but scientists were bitterly divided on it. Chemists knew that reactions had exact, quantifiable constraints, and this pointed to matter being made of discrete units. This underlies the 'balancing' of chemical equations, such as $6CO_2 + 6H_2O \rightarrow C_6H_{12}O_6 + 6O_2$: the photosynthetic production of glucose and molecular oxygen from carbon dioxide and water. But many thought these integer relationships were simply a book-keeping tool, and that matter remained continuous and smooth 'all the way down'.

Speaking of the world's most famous molecule, it is not widely appreciated that the well-known composition of water, as H_2O, was determined in 1804 by Joseph Louis Gay-Lussac and Alexander von Humboldt; and that the latter was perhaps the most prolific amalgamation of scientist, naturalist, explorer and writer that has ever lived.[14] Most of Humboldt's natural history work occurred in South America, where the Humboldt current flows and the Humboldt Penguin *Spheniscus humboldti* lives. He completed accounts for over 100 animal species; nearly 300 plant species bear his name, and more places around the world are named in his honor than for any other individual. (Upon joining the union in 1864, the state of Nevada was very nearly named Humboldt.) 'Everything is interconnected and interdependent,' he wrote, influencing Henry David Thoreau and other naturalists. With all that, few today recognize his name or know of his exploits.

This is all we will say about the unlikely story of Lewis, Clark, Brown and Einstein, and the initial evidence for molecules. But we haven't seen the last of Brownian motion, for it has a universality that is not confined to the microscope slide. In the next chapter, we'll move to the oceans, where the movements of seabirds will re-enact a similar pageant, and further atomic physics will help account for the ways pelagic species survive in their harsh environment.

Chapter 3

On the Open Seas: Length Scales, Migration and Molecules

Though our planet is mostly ocean, a pelagic birding trip can be an otherworldly experience. The birds are extraordinary, the identifications challenging, and there is a visceral sense of sheer vastness. In this chapter, we'll consider the movements of seabirds and grapple with the scale of migration. We'll also zoom in to the molecular level, studying pelagic birds from the inside to reveal how they survive in such harsh conditions.

Migration and large scales

Arctic Terns *Sterna paradisaea* are birds of extremes. They experience more daylight per year than another other creature on earth, and would warrant the name 'sunbird', had that moniker not been already taken.[1] They chase sunlight in a pole-to-pole campaign, in lockstep with the seasons; the course of their lives prescribed by the planet's axial tilt.

Those Arctic Terns that breed in the Netherlands take circuitous routes along the west coast of Africa, eastward to Tasmania, and south to Antarctica. By the time they cycle back, they might have racked up 90,000 kilometers, making their migrations the longest among all animals.[2] Given how they split their time between hemispheres, we might deem their name a northern chauvinism. But they are all hatched as boreal natives, which confers arctic citizenship, as it were.

We have no native grasp of the distances these birds cover. Our coarse senses expose us to a limited portion of the world, making it difficult to comprehend even a fraction of the earth's size; much less its entirety. Some analogies, and a bit of visualization practice, are needed for us to improve our spatial intuition.

We might start with the mental image of a meterstick, the standard of measure nearest to our own size. Going up an order of magnitude—that is, multiplying by a factor of ten—we have the width of three lanes on the highway. A hundred meters is the size of a football field, a distance readily taken in at any high-school campus. Scale it up again to reach the kilometer, and now visualizing distances becomes more challenging. Finding some shared reference that we can point to isn't so easy. The towers of the Golden Gate bridge are just over a kilometer apart. An airport runway is 3 to 4 kilometers long. At the seashore or on the plains, the horizon line lies about 4.5 kilometers away. That these examples become less insightful illustrates how a sense for increasing scale only gets harder. Some other way to grapple with it would be helpful.

We often mark off large distances by considering travel time. A brisk walking pace makes a kilometer go by in about 12 minutes, and two hours of walking would cover 10 kilometers. Driving for an hour at highway speeds takes us 100 kilometers. For 1,000 kilometers—roughly the distance from New York to Chicago by air—we might imagine several hours on a plane, but flying speeds are hard to intuit, and our calibrations are less trustworthy. Ten such flights would span equator-to-pole, a journey of some 12 hours by commercial jet. If we can imagine the distance covered during 50 long hours on a plane, we'll have a feel for half the tern's round trip. Otherwise, we could dedicate 12 hours a day to highway driving or walking, and cover that distance in four weeks or just over two years, respectively.

In zooming out from the meterstick to the equator-to-pole distance, the scale changes by *seven orders of magnitude*: that is, by a factor of 10,000,000, a figure needing seven zeros. A different tack might help us better intuit this range; we can use smaller lengths and make analogies. Dividing the meter into ten parts, we obtain a span of about 4 inches, the size of a kinglet. Another such reduction yields a centimeter, the radius of a penny. Ten times smaller is the millimeter, half the thickness of the same coin. A tenth of a millimeter isn't hard to visualize; we can scrutinize and divide by ten the smallest spacings on a ruler, or note that it is roughly the width of a hair or the thickness of paper. Somewhere below that, our eyesight can do no more. We'll content ourselves with a mental picture of a hair's width divided into ten equal parts and stop there. This distance, a hundredth of a millimeter, lies just beyond the native resolution of our eyes, but not our imagination.

Stepping back, let's agree that a nominal range that we can readily visualize stretches from a tenth of a hair's breadth up to a football field. That's seven orders of magnitude. If we can imagine scaling ourselves down to the low end of that, we'll grasp the degree to which the earth's size dwarfs us.

Big numbers are difficult to grapple with, whether they refer to distances or to something else. To gain some perspective on large quantities, suppose that we begin counting the seconds as they tick away. We'd need just under

17 minutes to reach 1,000. To reach one million, we'll have to count continuously, night and day, for over 11 days. To count to one billion, another number that simply 'sounds big' to most of us, would require almost *32 years* to reach at this rate. As for one trillion, our scheme completely fails, because one trillion seconds corresponds to some 32,000 years, a span beyond reckoning. Even if we counted off tenths of a second, then at the end of a typical human lifetime, only about 23 billion such little time units would have elapsed; leaving us with a number still *far* smaller than one trillion. The next 43 generations would have to keep up the effort to get there. Are such numbers 'too big' to have practical human relevance? Sadly, no. At the time of this writing, the United States national debt stands at over 35 trillion dollars.[3]

Today we are fortunate to have other tools that can help us develop a feel for changing scales. With a smartphone alone we can command a virtual Earth to move under our fingertips, ready for inspection at dizzying levels of detail. The Google Earth application, for instance, lets us see the world from a range of distances: from close enough to resolve objects only 10 centimeters in extent, out to a view from 63,170 kilometers away, from where the entire globe appears to fit in one's hand.[4]

Seeing the world from these novel perspectives reveals a beauty that is inaccessible to us otherwise, such as the fractal composition of some geological features. An intricate self-similarity is evident—like that which emerges in abstract objects such as the Mandelbrot set from Chapter 1—and is on display all over the globe, with the patterns emerging as the level of magnification changes. In river deltas, the channels split into ever smaller branches; jagged coasts are chiseled into nested corrugations; vermiculite ridges and rows of sand dunes go on like nested Russian dolls.[5]

All of these exercises will likely make us feel duly minuscule but, in reality, we inhabit a middle realm. We are dwarfed by the scale of the planet, yet are simultaneously gargantuan in relation to the building blocks of nature. We've seen how the jiggling of microscopic objects confirmed that everything is made from granular units, in the form of molecules and atoms. This is an appropriate moment to take another look into that realm, further exercising our skill for intuiting distances.

Molecules and small scales

It is a lovely paradox that the appearance of everything around us, the details of *what* we see and *how* we see it, depend on things too small for us to directly perceive. For example, the varied blues of a jay are immediate and clear, but the micro-structure in the feathers that causes these colors is beyond our coarse senses. We can discern it only through a process of expanding our perceptual range via experiment, deduction and clever engineering. (We will cover this in detail in Chapter 8.)

Our innate sense of scale might function down to fractions of a millimeter, but to dive into smaller realms, we'll need to leverage the method we just used to scale up. Suppose we start with a small droplet of pond water—about a millimeter in diameter—and imagine it magically swelled up by a factor of 100,000, making it as large as a football field. In such a blown-up world, the thickness of a sheet of paper, nominally about 0.1 millimeters, would correspond to 10 meters, the width of an end zone. If we now hold our hands 10 centimeters apart, we have a sense for the size of a micron (or micrometer), which corresponds to one thousandth of a millimeter. The micron is a ubiquitous and convenient unit of measure for working in the sub-millimeter realm.

Most pond water teems with single-celled microorganisms. Those of us unfamiliar with this part of the living world might be tempted to think of all such creatures as 'microscopic' in a uniform kind of way. But even considering only one category of life, the various forms of planktonic algae, their *relative* range in size is as broad as that for all the plants found in a tropical rainforest.[6] In our expanded droplet-as-football-field model, the largest protozoa—such as amoebas and paramecia—would be 20-meter-long leviathans (their nominal size can be up to 200 microns). Their smallest cousins, the bacteria and algae on the order of 0.2 microns, would appear on our field to be no larger than coins.

The *Clarkia* organelles that shifted about under Robert Brown's peering eye reside near the low end of this organic length scale. On our field, the bulkiest ones would take on the size of a soccer ball. Speaking of Brownian motion, one might wonder why almost eight decades passed before the jittery movement was explained by Einstein. After all, optical instruments steadily improved in quality throughout the 19th century; why didn't we eventually achieve the magnification that allowed us to see the water molecules?

The problem is that any optical device is inherently limited by the very light it uses, and light has its own size, a particular spatial extent, that is far larger than most molecules. This is a fundamental fact that we will return to again. Paradoxically, light by its very nature shields the very small from our eyes and our optics, and it took much time and cleverness to break through this barrier.

Recall that light is a wave in the electromagnetic field, a sinusoidal ripple of crests and troughs. Its wavelength—the distance from one crest to the next—can range from about 0.38 microns (for violet light) to 0.7 microns (for red). On our field, these correspond roughly to the sizes of ping-pong and tennis balls, respectively. Anything smaller—and water molecules are *much* smaller than 0.38 microns—will be impossible to make out in the blur obtained by wielding too coarse a tool.

We can think of it with this analogy: we see Brown's organelles via a process that is akin to shooting ping-pong balls at a soccer ball. We could subtly change our aim and map out the directions at which our projectiles bounce off, and from this, deduce the size and shape of the larger ball. But attempting to use light to see a molecule would be akin to aiming our ping-pong balls at a grain of sand. Such an effort could reveal nothing about the size, shape or structure of the minuscule target. There is no way around this fundamental limitation of light due to its (relatively) huge wavelength. Only in the 20th century would we develop other tools, such as electron microscopy and atomic force microscopy, that could probe down to the molecular level.[7]

Let's now start heading toward that molecular level using our droplet and football field model. Just as we used a factor of 1,000 to go from a millimeter to a micron, we'll divide the micron by the same amount to arrive at the *nanometer*. Several hundred nanometers (tenths of microns) correspond to the wavelength of visible light and the size of the largest viruses. Tens of nanometers measure the smallest viruses, some hefty protein molecules, and the scale at which much of our technology can now operate—although this is ever-changing. In 1984, the smallest feature in a semiconductor device was about 1 micron in extent. In 2016, it was ten nanometers, 100 times smaller. The steady growth in computing power and digital camera performance that we take for granted continues, in part, because we keep shrinking these components, allowing us to put more of them into the same volume.[8] We'll look into semiconductor nanotechnology in Chapter 9.

Within our field-sized droplet, the smallest of viruses and the features of modern transistors would appear a few millimeters in extent. Ten times smaller, at a hair's width, would be our expanded nanometer. Here, we'd find various organic molecules such as the nucleobases that connect up into DNA, or sugars like glucose. These are usually built from a dozen or more atoms, indicating that simple molecules and individual atoms themselves occupy a regime of tenths of nanometers. We are back to splitting hairs: a tenth of a hair's width on our field marks the sizes of the fundamental components of chemistry.

Note that, coincidentally, we must scale down by the same proportion to reach atomic sizes as we must scale up to reach earth-size dimensions. This was not lost on Richard Feynman, who pointed out a clever way to never forget the ratios: if we imagine an apple, and it swells up in size until it becomes as large as the earth, then the atoms inside that apple will now be as large as the original apple was.[9]

Having worked our way down to the molecular level of water, we will stay here to consider how seabirds survive in a harsh environment that is deadly to terrestrial life.

Seabirds and saltwater

Water is a primary component of all living things, making up most of the liquid (cytoplasm) within cells. The cytoplasm and other cell contents, such as the nucleus and mitochondria, are enclosed by a membrane, which protects the cell and controls what enters and exits. Some of this regulation is done by the membrane itself, which can act as a selective gate because it is *semi-permeable*; that is, certain molecules can pass through it by the process of *diffusion*. Larger molecules and ions (which possess a non-zero electric charge) cannot use diffusion to get in or out of the cell unless there are *channels* embedded in the membrane that permit it.[10] We will discuss these channels momentarily.

Let's look closely at diffusion. When a molecule on either side of the cell membrane collides with it, there are two possible results. One is to recoil off, transferring momentum and thereby causing *pressure* against the membrane. If the liquid compositions are similar inside and out, the total pressure—the collective force per area from countless collisions—will be the same on the inside as on the outside. The other outcome is that the molecule will pass through the membrane. This will occur at the same average rate—if the liquid compositions are the same—on both sides, so there will be no *net transport* in or out.

If, however, one side of the membrane has a greater concentration of solutes (substances mixed into the water, such as ions) that cannot pass through, this asymmetry can have a marked effect. Suppose that the liquid inside the cell is highly saline—containing large amounts of sodium and chloride ions—but that the water outside is less so. On the inside, both water molecules and ions exert an outward pressure on the membrane. The presence of the ions means there are fewer water molecules striking the inner surface that could potentially pass through it. Since there are fewer ions outside the membrane, there are proportionately more water molecules colliding against it, and hence more that can *move in*. As a result, water will preferentially enter a more saline cell, causing it to swell and perhaps burst.

In the reverse case, when more ions happen to lie outside the membrane, water will exit the cell, causing it to lose volume and contract. Exposing cells to a sufficiently saline environment thus leads to *dehydration*—the paradoxical result of taking in too much seawater.

The cells of seabirds and other marine animals are as prone to this kind of dehydration as ours. Seabirds solve the problem with specialized salt glands

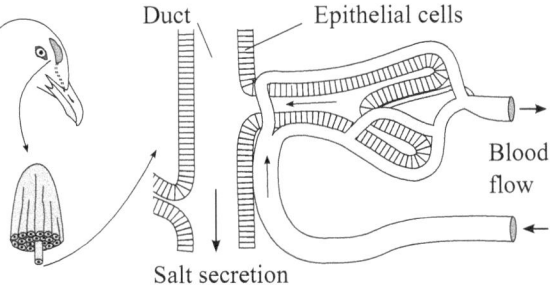

Figure 3.1 The avian salt glands are located over the eyes, with a central duct leading to openings on the bill. Each gland contains a series of parallel ducts, each of which branches into smaller ducts surrounded by specialized epithelial cells. A network of capillaries carries blood past these cells, which will extract the salt and pass a saline solution to be expelled. [Modified from Fánge, Schmidt-Nielsen and Osaki (1958).[11]]

located above their bills that continually remove the excess sodium and chloride ions that have gotten into the bloodstream. This clever chemical engineering is well worth exploring.[12]

Let's start with the large-scale operation of the gland, where salt is extracted from the blood supply and expelled in a concentrated solution. A schematic of the structure is shown in Figure 3.1. Capillaries move the blood past a network of hollow tubes made of specialized epithelial cells. The 'basal' sides of these cells are exposed to this blood flow, while their opposite, 'apical' sides form the interior (the 'lumen') of a duct. Various ducts splice with others, eventually forming a main duct. The ducts will carry a salt-rich solution extracted from the blood, which is eventually excreted from the body via openings on the bill.

The plumbing of this machine is rather straightforward. The challenging part is the work that the epithelial cells must do to extract the ions from the blood and move it into the lumen of the duct. To understand how this job is done requires some familiarity with a few mechanisms that cells employ in the business of transport: the movement of molecules across the cell barrier by means other than simple diffusion. This is where we need the channels alluded to above.

Facilitated diffusion channels are structures that allow certain molecules to move through of their own accord, which is what occurs if the *concentration gradient* is favorable. In other words, if there is a higher concentration of ions outside the cell, they will flow through the channel until the concentration inside is the same. Examples include chloride channels that transport the Cl⁻ ion; potassium channels to move K⁺ ions; and aquaporins, which as the name suggests, facilitate a greater flow of water than the cell membrane can support.

Then there are *active transporters*, which, unlike the diffusion channels, can move molecules *against their gradient*, from lower to higher concentration.

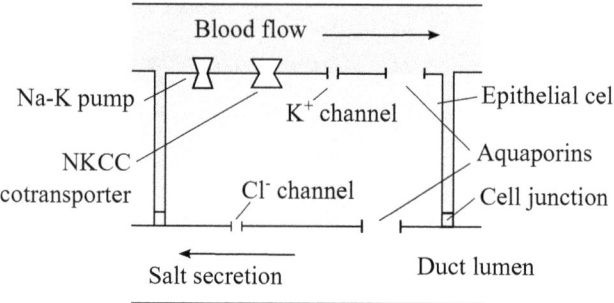

Figure 3.2 Schematic of an epithelial cell in the salt gland. The basal surface on the top borders a capillary carrying salt-rich blood. The membrane here contains facilitated diffusion channels (K^+ and aquaporins) as well as active *transporters* (Na-K pump and NKCC cotransporter). The apical surface on the bottom borders the lumen of the duct, which carries off the extracted saline solution. The membrane here includes aquaporins and Cl^- channels. Cell junctions connect the apical surfaces of adjacent cells. [Modified from Braun (2015).]

This comes at a cost, in that the cell will have to use some potential energy stored in the form of the adenosine triphosphate (ATP) molecules produced by mitochondria, which are the 'baseline currency' of cellular energy. The active transporter of most interest for us here is the *sodium-potassium (Na-K) pump*, which 'spends' one ATP molecule to cycle three Na^+ out of the cell and bring two K^+ in. Another example is the NKCC *cotransporter*, which uses the Na+ gradient created by the Na-K pump as an energy source to allow simultaneous movement of Na^+, K^+ and Cl^- into the cell, the latter two against their concentration gradient.

The channels and transporters we have mentioned, together with judiciously arranged epithelial cells within the network of capillaries and ducts, are all the components needed to make the filtering system that marine birds use. How it all works is most easily understood if we break the system up into steps, keeping in mind that, in execution, every process is running concurrently.

Figure 3.2 shows a simplified view of an epithelial cell, which is a close-up of the structure shown in Figure 3.1. On the top is the basal surface, which interfaces with the blood flow carrying the excess salt. On the bottom, the apical surface of the cell borders the lumen of the duct, into which the salt will pass. Each cell has neighbors pressed up against it on the sides, but importantly the boundaries, or cell junctions, will permit some molecules to move in the gaps between them.

Various transport channels are also indicated. As this is a schematic, only one of each is shown, whereas real cells will have multiple channels. There are other membrane features that we are ignoring here as well. What is critical is how the channels are arranged and used. Both cell surfaces contain aquaporins

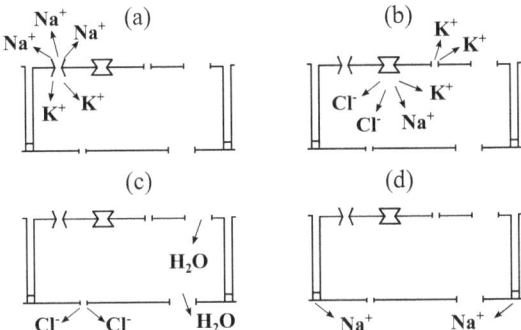

Figure 3.3 Key steps in the extraction of salt. (a) The Na-K pump moves three Na⁺ out of the cell and two K⁺ in. (b) The K⁺ ions are free to return back outside the cell via channels, and due to the lowered Na⁺ content inside, the passive cotransporter allows Na⁺, K⁺, and Cl⁻ to flow in. (c) The excess Cl⁻ in the cell moves out through the Cl⁻ channel into the duct, which also draws water through the aquaporins. At this point, Cl⁻ has been moved from the blood into the lumen, while the Na⁺ is more concentrated on the basal side. (d) Excess negative charge in the duct attracts the Na⁺ through junctions, completing the process. [Modified from Braun (2015).]

to facilitate the flow of water. Cl⁻ channels appear on the apical side, while the basal side has passive K⁺ channels and NKCC cotransporter; and the workhorse, the Na-K pump.

Figure 3.3 breaks out various steps in the process. In (a) the energy stored in an ATP molecule is used to operate the Na-K pump, which moves three Na⁺ out of the cell and two K⁺ in. The K⁺ ions are free to return back outside the cell via channels, and due to the lowered Na⁺ content inside, the passive cotransporter allows Na⁺, K⁺ and Cl⁻ to flow in, as shown in (b). The buildup of Cl⁻ in the cell favors it moving out through the Cl⁻ channel into the duct, which also draws water through the aquaporins, in (c). At this point, the work done by the ATP has simply drawn the Cl⁻ out of the blood and into the lumen, while the Na⁺ has now become more concentrated on the basal side. This is where the interfaces between the cells play a role. As shown in (d), the excess negative charge in the duct attracts the Na⁺ ions (thanks to the electrostatic force) in through these junctions, where they can pass into the lumen, completing the process.

For the ATP 'fuel' to power this process, some of its chemical potential energy must be spent and leveraged into the mechanical energy of operating the Na-K pump. Spending that energy removes one of the three phosphates, leaving a different molecule—adenosine diphosphate (ADP)—behind. Like a spent cartridge, ADP can be reloaded when the cell utilizes energy from glucose or fats—other powerhouse molecules—to reattach the phosphate.[13] These fuels derive their potential energy ultimately from sunlight harvested by the plants at the far end of the nourishment supply chain.

Without this energy, the entire salt-extraction scheme breaks down. Left to themselves, the ions would only move *down gradient*—that is, from regions of high to low concentration, in the process of diffusion, the inevitable result of the same jostling motion of the molecules that drove the Brownian motion introduced in Chapter 2. A drop of ink will diffuse through a glass of water even without stirring, becoming evenly distributed without any 'master plan' to equalize through the mixture. It is merely the inevitable result of collected, random movements, as would be the spread of salt throughout a seabird's cells without the action of glands to pump it back out. But the movements described by diffusion aren't just something that molecules and atoms do. It occurs across the entire range of length scales that we've been studying. It can be manifested in the movements of birds themselves, which we will begin to investigate in the following section.

Random walks and albatross

1905 was a momentous year, with Einstein publishing his relativity, quantum physics and Brownian movement papers. It was also when statistician Karl Pearson introduced the concept of the *random walk*.[14] Studying the dispersal of a swarm of mosquitoes, Pearson modeled individual insects moving via short flights in random directions. Sometimes termed a *drunkard's walk*, this motion might emulate an inebriated person moving in some direction, stopping (or falling) before staggering off randomly in another direction, and so on. Every leg of this 'journey' is unrelated to previous ones, and we might expect the drunk to go nowhere. The reality, as we will see, is far more interesting.

The path traced out by a microscopic particle jostled about by water molecules during Brownian motion is also a random walk. So too are stock prices, the movements of beetles, butterflies, humans and elk, the pathophysiology of a cancer cell invasion, swarming behaviors and the flow of traffic.[15] The process of diffusion, the relentless flow of a substance from high to low concentration, is the inevitable, collected outcome of each bit of that material performing its own random walk. The random walk is a universal process that warrants a closer look.

Figure 3.4 shows several random walk examples. In (a) the path is based on 1,000 steps using a simple version where the steps are limited to the four cardinal directions, selected randomly, with the distance moved always remaining the same. If we allow direction and step lengths to vary randomly as well, we obtain the path shown in (b), also using 1,000 runs. Let either process continue and we'll produce images full of detail—as shown in (c) and

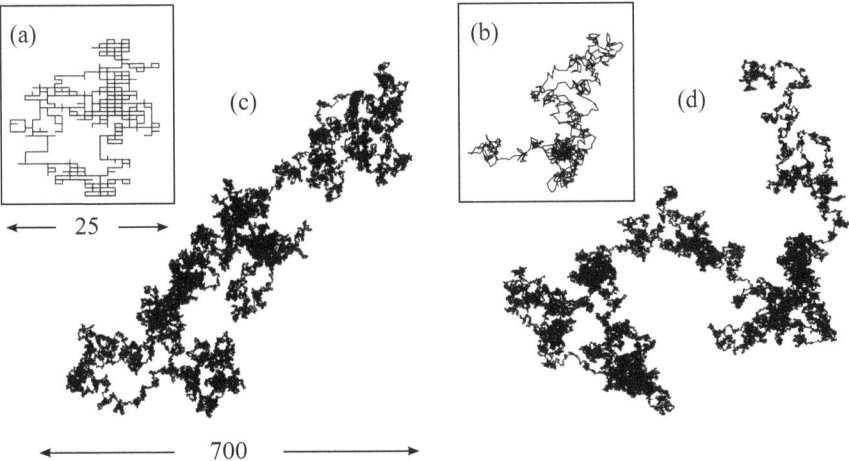

Figure 3.4 Random walk examples. (a) Based on 1,000 steps, using a simple version where the steps are limited to the four cardinal directions, each having a length of one. (b) Another run of 1,000 but with variable direction and step size. (c) The process from (a) allowed to continue for 1,000,000 runs. (d) The process from (b) allowed to continue for 1,000,000 runs. The length scales for (a) and (b) are the same, as are those for (c) and (d).

(d)—each made with one million runs using the limited and free-ranging approaches of (a) and (b), respectively.

If the images in (c) and (d) remind us of fractals, it is because they are. Iterating a simple rule, even if it involves randomness, this again leads to designs of great intricacy. The fractal character only breaks down at small length scales, such as in (a) and (b), where the self-similarity stops. But as a walk proceeds, we must zoom out (note the change in scale) so as to take in its entirety, and the difference in the two schemes becomes too small to be resolved. It is a universal property of random walks that the *variance* of the position—the average of the squared distance from the start—must increase over time. The average distance will be zero, since there will be just as many negative as positive numbers. Squared distance, however, is always greater than zero, so the average will be positive.

It might seem that random walks should not be relevant to animal migrations, which involve a goal-directed kind of motion. However, persistent movement in a given direction—but with small random effects—can be accounted for with a *correlated random walk* model, such as the *Brownian bridge*. Unlike the typical walk, which moves off unconstrained, a Brownian bridge involves randomness between fixed start and end points. It is a natural choice for modeling bird migration, where the wintering and breeding grounds are set, but the specific routes are not. The Brownian bridge approach has been used to analyze sparse telemetry data and to help understand the movements of

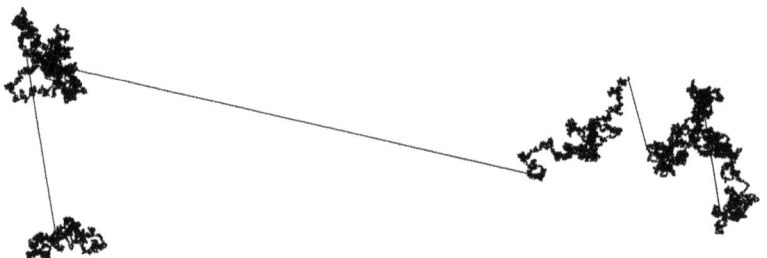

Figure 3.5 A Lévy flight example, based on 50,000 steps.

Osprey *Pandion haliaetus*, several New World vulture species, and the flyways used by Asian waterfowl species.[16]

There is another version of the random walk that once had a controversial relationship with the subject of bird movements. It goes by a name conferred to it by Benoit Mandelbrot, whom we met in Chapter 1, to honor the mathematician that he studied under, Paul Lévy. It is called a 'Lévy flight' or a 'Lévy walk'.[17]

With a typical random walk, the step lengths can be generated in any random way, provided the lengths are capped by some maximum value. That is, the distribution must be 'short-tailed'. If this requirement is not met, if the numbers are drawn from a 'long-tailed' distribution where the occasional *big number* shows up, the results are very different. This is what a Lévy flight is. When mapped out, it looks like different random walks that are connected by long, linear 'flights' between them. These are of course the huge steps that show up only occasionally: the points from the long (or 'heavy') tail. An example is shown in Figure 3.5.

It is generally thought that the most effective foraging behavior under certain conditions—such as when prey is rare and slow-moving, and the forager has limited knowledge of its whereabouts—is the Lévy flight.[18] Starting in 1996, researchers began looking for evidence of animal behavior that exploited this tactic. The work that kick-started the study of Lévy flights in ornithology involved the movements of the Snowy (previously 'Wandering') Albatross *Diomedea exulans*.[19] A physics graduate student at Boston University, Gandhimohan Viswanathan, analyzed data from monitors placed on birds that indicated when they were on water. His results showed excellent agreement with the expected distribution of step sizes in a Lévy flight. However, when other researchers failed to replicate the results, Viswanathan's team realized that the sensor data were not always interpreted correctly: what were thought to be long flights were sometimes the result of the birds on dry land, attending their nests. By 2008, the question of the Lévy flight use by animals was undecided.[20]

With more data collection and new studies under different conditions, the evidence for Lévy flights began to accumulate again, and the albatross became the subject of renewed attention. Detailed tracking of movements over shallow water versus deeper water showed different patterns, with Lévy flights occurring in the latter case. Among other birds, Black-browed Albatross *Thalassarche melanophris*, Cory's and Scopoli's Shearwaters *Calonectris borealis* and *C. diomedea* and Magellanic Penguins *Spheniscus magellanicus* have also been reported to use them.[21] Moreover, turtles, tuna, *E. coli* bacteria, sharks, mussels, mud snails, honeybees and even human hunter–gatherers have been the subjects of studies that point to the same underlying foraging strategy.[22]

Lévy flights were understood to be efficient search methods before researchers demonstrated that they were actually employed by organisms. A reverse approach is to look at biological behaviors and ask if they provide efficient ways to solve optimization problems. One example is a problem-solving program that emulates the behavior of a colony of ants.[23] The strategy is to run parallel and independent searches for solutions, with periodic exchanges of information between the ants or 'agents' in order to select the best option and continue looking for improvements from there.

Tuning up our own algorithms based on insights from animal behavior can be useful, and it is a busy enterprise, although not without controversy.[24] Critics point out that the plethora of 'novel' approaches—such as those based on bees, mayflies, termites, fruit flies, glow worms, fireflies and others—differ only in marginal ways from the original approaches, such as that used in the ant colony algorithm.

Avian behavior alone has led to 'bird swarm optimization' techniques, often used in financial analyses, an 'emperor penguin colony' scheme that emulates huddling behavior employed to share body heat, a 'cuckoo search' algorithm based on nest parasitism, an 'eagle strategy' that purports to combine Lévy flights and firefly behavior, and even approaches based specifically on Golden Eagles *Aquila chrysaetos* and Harris's Hawks *Parabuteo unicinctus*.[25] Whether or not these provide useful techniques for humans to solve their problems, there appear to be countless ways that biological systems can cross-fertilize with seemingly unrelated fields. In the following chapter, we will look more closely at bird swarms, and see even more examples of random walks in the avian world. And we'll explore the molecular mechanism that makes world-spanning migration feasible.

Chapter 4

In a City Park: Movement, Murmuration and Magnetism

City parks can be bird magnets, veritable oases dotting concrete deserts, and fitting places to ruminate on the movement and migration of land birds. In this chapter, we'll locate ourselves in Manhattan's Central Park, launching site of a legendary species invasion, and the place where a Nobel-winning physicist who revolutionized our understanding of matter became a devout birder as a child. We'll see how random walks explain the expanding ranges of birds and the origins of animal patterns, a mystery solved by the genius who broke the 'enigma code' in World War II. And we'll explore how migrant birds sense the earth's magnetic field.

Patterns and emergence

The story of the introduction of European Starlings *Sturnus vulgaris* in the Western Hemisphere is part of birding folklore. It is widely held that one Eugene Schieffelin was responsible for setting some 60 birds loose in New York's Central Park in 1890, because he hoped to see every bird mentioned by Shakespeare living in America.[1] While the release undoubtedly took place, a recent study has concluded that the Shakespearean angle was wholly fabricated, as was the idea that this was a one-time event.[2] Similar releases of starlings had been going on throughout the continent for decades prior.

We've learned to mitigate the disruptive effects of these exotics on native ecosystems, at the cost of the additional labor needed to protect native cavity nesters such as Eastern Bluebirds *Sialis sialis*. Yet there is much to admire in these birds, even if we would rather enjoy them from European soil. Their singing and mimicking skills earned their mention in Shakespeare's *Henry IV, Part I*. Mozart kept one as a pet and remarked on its ability to pick up the opening melody of the third movement of his *Piano Concerto No. 17*.

Apparently quite fond of birds generally, the great composer even wrote verses for its funeral.[3]

Starlings are stunning to behold in their breeding plumage, with a rich network of whitish spots contrasting with their nominally glossy black coat. A close look reveals that this pattern arises simply from dark feathers developing lighter color on their tips. The periodic, shingled arrangements of the feathers—which vary smoothly in length and size around the body—inevitably produce a speckled outer presentation.

Spots and stripes adorn many animals and often such patterning cannot be accounted for so simply. Is there a tidy explanation for the formation of a jaguar's spots, or other examples of the 'pied beauty' that inspired Gerald Manley Hopkins to write, 'Glory be to God for dappled things … For rose-moles all in stipple upon trout that swim …'?[4] That question was posed and answered by an unlikely visionary, the pioneer of computer science who cracked the 'enigma code' during World War II, Alan Turing.[5] And it involves the same process governing diffusion that we've been finding everywhere.

In his last published paper, 'The Chemical Basis of Morphogenesis', Turing stepped outside of his familiar domains and into biochemistry. The article, which would go unnoticed for decades, described how quasi-repetitive patterns could arise from the interaction of two chemicals. One chemical, or agent (we'll call it '*A*') tends to activate, or promote, its own production, while the other ('*B*') acts to inhibit the process. Placed together, *A* and *B* become mixed as they each spread out via diffusion; and at any point in space, their local relative concentrations will determine how they react. The two agents will accumulate in distinct regions, where one or the other flourishes. And if they produce different colors, spotted or striped patterns naturally emerge. Diffusion moves the agents through space, reaction alters their abundances where they meet.

Turing's equation, also called the *reaction–diffusion equation*, captures this in terms of the *rates* at which the concentrations change. We'll refer to the respective concentrations of agents *A* and *B* as *a* and *b*, which vary with location and time. The rate at which *a* changes (written as $\partial a/\partial t$) is:

$$\frac{\partial a}{\partial t} = R(a, b) + D\nabla^2 a$$

A similar expression governs *B*.

The term *R(a,b)* governs the reaction: how the meeting of molecules affects the rate at which *a* changes. Its details needn't distract us; suffice to say that, for any two concentrations *a* and *b*, there is some rule *R* for how they react. The other term describes how *a* varies throughout space: the triangular symbol indicates the *gradient*—that is, the spatial change—in *a*. We may think of

Figure 4.1 Two examples of Turing patterns resulting from the reaction–diffusion equations.

gradient in terms of changes in density. A drop of ink added to water will initially appear tight and localized; its sharp boundary is precisely what is meant by 'large gradient'. As it diffuses and becomes less dense, the gradient becomes smaller. If we remove the R term, we are left with the classical diffusion equation, which states that the rate at which concentration changes is proportional to the square of its gradient. Dense, localized material spreads out quickly but the process slows and eventually stops once it is fully mixed.

We can solve such equations and plot the resulting patterns with just a few lines of programming code. But in 1952, Turing didn't have much computing power, and had to crunch it all out on paper. Genius that he was, he could yet see that it worked. In Figure 4.1, patterns generated using Turing's method are shown. The two runs differ only in the coefficient D that controls the diffusion, and the details of the reaction term R.[6]

Complex patterns in avian plumage can be accounted for using such equations.[7] Chevrons, paired spots, arrays of spots, bars, central patches and so on, can emerge if Turing's process occurs during feather growth. Several examples of patterned feathers and corresponding reaction–diffusion model results are shown in Figure 4.2. Moreover, it isn't just the distribution of pigmentation that can be explained in this way, but also the form of the feather itself. It has been demonstrated that specific proteins can perform the roles of activator and inhibitor during the embryonic feather development stage to create repeating, stripe-like structures that will separate into the feather barbs themselves.[8]

As with fractals, we see unexpected detail arise from simple rules. The equation might not have appeared simple at first glance, but knowing what the terms mean, we see that it is merely shorthand for a straightforward idea.

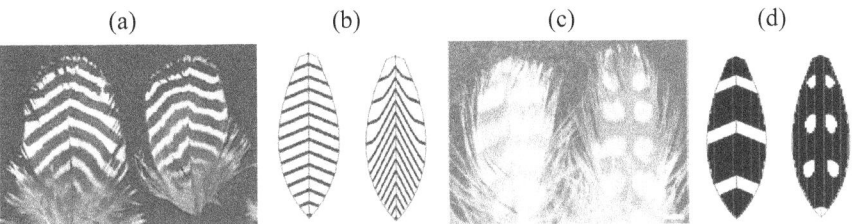

Figure 4.2 (a) Black Bustard *Eupodotis atra etoschae* feathers and (b) corresponding Turing pattern modeling results. (c) Crested Serpent Eagle *Spilornis cheela panayensis* feathers and (d) corresponding Turing pattern modeling results. [Modified with permission from Prum and Williamson (2002).]

Returning to the story that started this chapter, the expansion of the starling's range is itself a textbook example of a random walk, the process at the heart of diffusion. Individuals move off in irregular directions and reproduce, and their descendants do the same. Each heading and distance is unplanned, but in inevitable random walk fashion, a flow from higher to lower concentration must ensue. Without any directed goal, the variance will increase and the population will spread through space, governed by the same Turing equation with the reaction term removed.

Another captivating behavior of starlings is their proclivity to form huge, shifting flocks, or *murmurations*. Especially when several thousand or more birds are involved, the display may seem precisely choreographed, appearing like a kind of single-celled, shape-shifting organism sliding over the landscape. The individuals all seem to be on the same page, somehow. In a sense they are, but not in the way we might think. With some modeling, we can show that murmurations emerge without any overarching goal.

When we speak of modeling here, we mean simulating a collection of objects or 'agents' by treating each of them as simply as possible—often approximating them as points—and then specifying a procedure for how they move. In 1986, computer scientist Craig Reynolds published an influential paper showing how the flocking behaviors of birds could be simulated by treating them as a collection of individuals that obey simple behavioral rules.[9] Essentially a collection of particles, which he referred to as 'boids', the flock evolves over time due to each boid following prescriptions for separation, alignment and cohesion. We will consider each of these in turn shortly; the key idea is that the rules are *local*.

A local rule means that how a given boid moves will depend only on how its nearby neighbors are behaving. Each member continually adjusts its motion in compliance with the same laws as its local situation changes. There is an iterative flavor to this and, similar to what we saw with Barnsley ferns and Turing patterns, complex global phenomena can emerge in the process.

Figure 4.3 A 'boids' simulation result for 6,000 individuals after starting with random orientations and velocities.

The first boid rule, for separation, requires that individuals will not crowd too close and interfere with others. Think of this as collision avoidance. The second rule, for alignment, specifies that individuals will attempt to move in the same average direction as their neighbors. The third rule, for cohesion, ensures that boids stay together. This is implemented by each boid looking at the average position of its neighbors and tending to move that way, without violating either of the first two rules. It is easy to write computer code to simulate many boids obeying these rules. The approach is so effective that Reynolds' scheme is used to produce realistic flock and schooling animations in major films.[10] Figure 4.3 shows an example result from a boids simulation based on 6,000 individuals that began with a random distribution, but quickly formed a cohesive murmuration.

The complexity of a starling flock is an example of *emergence*. This refers to an action of a collective that isn't explicitly specified, but which occurs as its individual members follow their own rules. Simply put, the whole is greater than the sum of the parts: what emerges isn't explicitly specified by the underlying laws. American physicist Philip Anderson—a pioneer of the idea—expressed it well: 'The behavior of large and complex aggregates of elementary particles, it turns out, is not to be understood in terms of a simple extrapolation of the properties of a few particles.'[11] This is not to say those particle properties are unimportant. Rather, they are *necessary*, but not *sufficient* for understanding what many particles can do together. It's another way in which the complex hides within the simple, and makes them both more intriguing. Emergence helps counter our fears that the reductionism of physics must entail a cold worldview, seeing everything as 'mere' aggregates of atoms. Nothing is 'mere'.

Nobel laureate and birder

Another early champion of emergence was an explorer of nature at its smallest scale, the physicist who took the art of reduction further than it had ever gone. He was also, arguably, the smartest birder to ever lift a pair of binoculars.[12]

Murray Gell-Mann revolutionized our understanding of the subatomic realm, but his affinity for science began in Manhattan's Central Park where, as a precocious child, he learned to identify birds. Based on some of his later antics—wooing his wife by taking her to Scotland to view puffins, attempting to see half the world's species, and at least once forgetting to show up to speak at an international conference because he was out birding—we might infer which pursuit he held dearest. Early in his career, he had an outright disdain for physics, perceiving it as disconnected sub-fields lacking the beauty and unity he saw elsewhere. When he was done, he'd leave physics far more beautiful and unified.

During the height of World War II, Gell-Mann started his college career, entering Yale at age 14. Before the war, the inner workings of atoms were becoming clearer, and the existence of their constituent parts—protons, electrons and neutrons—had been established by experiment. A few other particles, short-lived and seemingly unrelated to everyday matter, had also been found, but the atomic constituents appeared to comprise the rock bottom of matter. A trio consisting of positive, negative and neutral particles had an elegant simplicity.

In the decades that followed, however, new particles kept being discovered. Some were lighter than electrons, others were heavier than neutrons. Most were unstable, transitory things that would quickly break up into other particles. By the 1960s, *hundreds* of new examples had been identified—such as muons, kaons, pions and so forth—comprising a 'particle zoo' that embarrassed physicists who expected nature to be more parsimonious at the subatomic scale.[13]

Looking to find order in the growing mess, Gell-Mann sought to establish a kind of taxonomy. Following Mendeleev, who had brilliantly deduced the underlying periodicity of the elements, Gell-Mann shuttled the particles around into geometrical patterns. And like Mendeleev, he found a sensible structure with a few gaps, where undiscovered pieces—if they existed—would make the puzzle complete. He boldly predicted they would be found, and they were. His realization—recognized with a Nobel Prize in 1969—was that all heavier particles, collectively called *hadrons*, could be accounted for if they were composites of several, more basic particles that he called *quarks*. This was a term coined by James Joyce and accorded various meanings, including the cry of a gull.[14]

Quarks warrant their unique name, for they have uncommon properties, such as fractional amounts of electric charge, something never before seen. Gell-Mann's first model specified three varieties: an 'up quark' with a charge of ⅔ that of a proton, a 'down quark' and a 'strange quark' each with a charge of $-\frac{1}{3}$. A proton consists of two ups and one down, and a neutron of one up and two downs; this neatly gives them total charges of one and zero, respectively. The strange quark was needed to explain the more esoteric, short-lived particles being discovered, and doesn't occur in 'everyday' matter.

For the scheme to work, quarks would also need to possess an entirely different property, known as *color*, or *color charge*. It is an unfortunate name, as it has nothing to do with the common meaning of the word, but there is a logic to it. Recall that electrical charge comes in two varieties, so that we need an equal amount of positive and negative to achieve an uncharged, or neutral, result. For quarks, nature has gone one step further, specifying that three varieties are needed for balance. Hence, we call them blue, red and green because, when mixed, they achieve a neutral white.

The three quarks inside a proton or neutron each carry a different color charge, making the resulting particle 'colorless'. This is why we do not observe color charge on any particle we normally interact with. And quarks cannot be extracted from protons or neutrons, and cannot roam about as free particles. The force binding them together is so strong that if you do provide enough energy to pull them apart, that energy converts into the mass of new quarks popping into existence, forming new colorless composites.

Just as electric charge is tied to the electromagnetic force, color charge couples the aptly named *strong nuclear force*. It is a fundamental force, like gravity and electromagnetism, intrinsic to nature. But unlike these other two forces, which manifest in action over large distances, it has a short range, extending only over extremely small length scales about 100,000 times smaller than the size of an atom. In addition to keeping the quarks tied up inside protons and neutrons, the strong force acts between these particles, also known as nucleons, holding the entire nucleus together. Without it, the only element would be hydrogen; all the others have multiple protons in their nuclei, which would blast apart from the extreme electrostatic repulsion if the strong force was not there to compensate. This is also why nuclei need neutrons, as they supply additional strong force binding without adding more repulsive positive charge to the mix.

When we attempted to visualize atomic length scales in Chapter 3, we had to journey inward seven orders of magnitude smaller than a millimeter. To intuit the sizes of nuclei and nucleons, we must go in another five orders; a factor of 100,000. Expanding a large atom, such as uranium, to the size of a football field, we would find the nucleus to be about a centimeter across, while a single proton or neutron would measure around a millimeter.

So with all of that, we are again left with a mere trio of particles that can account for all everyday, stable matter: the electron, the up quark and the down quark. We needn't get sidetracked with the handful of other particles here; suffice to say that there are six quarks and six *leptons* (the category to which the electron belongs) that fill out the equivalent of the periodic table for matter particles. Given the name, we tacitly think of particles as tiny 'points': little bits of 'stuff' that each bear a minuscule mass and charge, standing in utter contrast to the expansive fields that carry the forces between them. This is incorrect, however; reality is far more elegant. If we look more closely at these connections, we'll get a better insight into the atomic structure of matter and, later, the navigational sensing of migrating birds.

Particles and fields

Birds within a species are often indistinguishable in the field. Tracking them as individuals usually takes great effort, such as netting them and putting bands on their legs. But this isn't always the case, as illustrated by one of Gell-Mann's birding stories.[15] In 1956, while driving with his wife outside of Los Angeles, he glimpsed a large, low-flying bird disappear over a hill, and suspected it might be a long-sought, rare endemic of the area. They pulled over and ascended the hill to look; his flannel suit and her skirt and heels soon becoming covered in thick, red mud. Eventually, they found a dead calf and 11 California Condors *Gymnogyps californianus* feeding on it. Gell-Mann realized that he was seeing a significant fraction of the entire species, decimated in number by human activity. But what struck him the most was how easily each individual could be told apart. Differences in feather loss and molt stage gave them away.

Subatomic particles—Gell-Mann's other preoccupation—are more like the members of a distant starling flock, because individuals cannot be told apart. Electrons have no equivalent to feather wear or molt, and we cannot capture and tag them. But whereas distant starlings are *practically* indistinguishable, particles are *fundamentally* indistinguishable. Why this is the case, and the implications of it, lie at the heart of the simplicity and complexity in nature.

It should seem uncanny that every elementary particle of a given species is like every other, regardless of their individual histories. An electron that has bounced around for billions of years within the core of the sun is no worse for wear, and carries no scuff marks or scars to distinguish it from one that has been floating in empty space for the same time. Moreover, should we attempt to corral up several of them in near proximity—within an atom, for example—they will somehow 'sense' what their neighbors are doing, and will be forced to occupy different niches. This is famously expressed through the various rules concerning states, shells and orbitals: the seemingly arbitrary constraints that we had to learn in high-school chemistry, usually without learning about their elegant geometrical origins.

That particles must be indistinguishable and can somehow 'know' what states their neighbors are in becomes clear upon grasping one of the most beautiful revelations that modern physics has given us. Namely, *there are no particles, there are only fields*.[16] That is to say—to go along with the force-carrying fields we have already met, which convey gravity and electromagnetism—there exist various matter fields. Particles of a particular species are not minuscule bits of some ultimate building material, but rather localized waves in a particular fundamental field. Fields are all there is, and everything is made from fields.

There is a single *electron field* that extends everywhere in the universe, and provided we bump it with the correct amount of energy, the localized wave disturbance that results *is* an electron. There is also an *up quark field*, a *down quark field* and so on, for every kind of particle we know of. These fields are strange things, fluid-like and yet granular in the sense that their waves are restricted to discrete levels, or *quanta*, of energy; hence their full designation as *quantum fields*. Each matter field extends through all space and is *coupled* to at least one of the fundamental force fields, such as the electromagnetic (EM) field.

To say that the fields are coupled, we mean that a ripple through one can potentially excite a wave in the other. For an analogy, consider the wind moving over a tallgrass prairie, coupling the air velocity to the height of the grass. Both of these things are fields, varying in time and place across the landscape. Meanwhile, air temperature also changes as clouds modulate the sunlight, but the temperature field doesn't affect (or couple to) the heights of the nodding stalks of grass. In a similar way, the quark fields are coupled to the *gluon field* (the carrier of the strong force), but the electron field is not. Electrons cannot 'feel' the strong force, no matter how intense it may be. At the same time, the electron and quark fields couple to the EM field. Charge can be seen as merely the property that couples a matter field to the EM field, and can be done in two ways (positive and negative), with the numerical value of the charge being a measure of how strong the coupling is.

This description of nature isn't merely an interpretation. These assertions make testable, quantifiable predictions. The dozen matter fields and several force fields, each coupled in varying degrees and following conservation laws that spring from underlying symmetries, all underpin the *Standard Model*: by far the most successful scientific theory ever developed. Richard Feynman was fond of pointing out that its predictions had been verified to an accuracy comparable to measuring the distance from New York to Los

Angeles down to the width of a hair. That was many decades ago. As of 2022, more detailed experiments have confirmed it 100-fold further: like measuring the circumference of the earth down to a tenth of the width of a sheet of paper.[17]

By understanding the primacy of fields, the subatomic world makes a bit more sense. Two electrons are not two different *things*, but two identical behaviors manifested in one single underlying entity. They cannot be 'tagged' and followed about, any more than the ripples on a rain-dappled pond passing through each other can be. The atomic rules (captured by the Pauli exclusion principle) that force electrons to inhabit different states, described in the language of shells and orbitals and so on, are a consequence of the field's innate properties. It isn't that a given electron can somehow 'know' what the others around it are up to. Rather, it's that the field cannot support two identical waves at the same spatial location; much like the rules of chess do not allow more than one piece per square.

The wave-like character of matter has huge practical ramifications. For example, the fact that electrons could maintain stable 'orbits' within an atom is otherwise inexplicable. Since the electrical force between electron and nucleus obeys the same inverse square law as gravity (leading to the iconic image of an atom with a nucleus as sun and electrons like planets in elliptical orbits), the electrons must be ever accelerating, just like the earth is ever falling toward the sun. But accelerating charges produce EM radiation, which would carry away energy, meaning that the electron should slow and quickly spiral into the nucleus.

This is not what happens, thanks to the wave nature of the electron. The first successful atomic theory recognized that electron orbits must have circumferences corresponding to one wavelength, or two wavelengths and so on, but nothing in between. This is inevitable when you place a wave in a circle so that it falls back on itself. Only certain circles are possible, with those supporting more 'waviness' corresponding to higher energies. This prevents any 'death spiral' because the electron can never move in a circle smaller than its wavelength.[18]

Molecules and atoms have appeared now in several chapters. This has been our first look at them from the inside, but it will not be the last. We will get to additional subatomic physics later—including more details of atomic and molecular orbitals—when we consider the origins of bird coloration and the principles of optical instruments. For now we'll focus on a particular interaction of molecular electrons in order to describe the recently discovered mechanism that is thought to underpin the physiology that migrating birds may use to navigate.

Cryptochrome and compasses

Having just explained matter as localized waves in fundamental fields, we must now acknowledge something rather difficult to understand in this picture, something that will send us right back to embrace our intuitive notion of particles. It is this: electrons (and quarks) are imbued with 'spin'; in some sense they *resemble* little spinning spheres, despite being nothing of the kind. Spin is simply an intrinsic property of the field, even if we cannot visualize how it transpires. So *if it helps to think* of electrons as tiny rotating objects, so be it. What is critical here is that every electron has two ways to spin; think clockwise or counter-clockwise. Moreover, while the persnickety field properties won't permit two electrons *in the same state* to overlap, two electrons spinning in opposite ways ('anti-parallel') *can* occupy the same space. Hence, another rule from high-school chemistry: that each atomic orbital may hold two oppositely spinning electrons.

The importance of spin is impossible to overstate. The pairing of electrons makes for such a stable, favorable configuration for the field, that it compels atoms to bond and form molecules. It also directly imbues electrons and quarks with a fundamental bit of magnetism. A 'clockwise' electron accords with its north magnetic pole pointing up. The opposite spin points down. The anti-parallel requirement for pairs corresponds to the tight embrace of two nearby magnets when each north is proximal to the other's south.

Like little compass needles, single electrons and quarks—we will refer to them simply as 'spins'—can rotate and point this way or that. A compass needle will try to align with whatever magnetic field is present: that of the earth, or a current-carrying wire, for example. A spin responds in a somewhat different way, by wobbling like a spinning top, its orientation circling around the field. This is *precessional* motion, illustrated in Figure 4.4.

A top might precess roughly once per second, but the wobbling of an electron is typically a million times faster, a rate in the range of *radio frequencies*

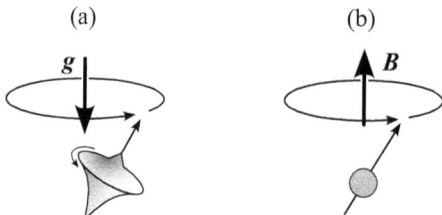

Figure 4.4 (a) A spinning top in the presence of a gravitational field *g* will precess in a circular manner around the direction of the field. (b) An electron in the presence of a magnetic field *B* will precess in the same manner.

(RF). The term RF has stuck with us ever since the first radio broadcasts, which utilized electromagnetic radiation in the megahertz (one million oscillations per second) band. The exact frequency is proportional to the strength of the magnetic field that the spin is exposed to; double the field and the precession rate doubles.

These effects might seem esoteric, mere curiosities of interest only to a few scientists. This is not the case. The precession of unpaired spins endemic to certain atomic nuclei makes magnetic resonance imaging (MRI) technology possible. Here, a body is exposed to both a magnetic field and RF radiation that stimulates the spins to precess. At locations where the magnetic field strength is appropriately 'tuned' with the RF frequency, *resonance* will ensue; the spins will readily absorb the RF energy, just as a child on a swing will move in huge excursions if we push at the correct rate. This permits a kind of triangulation and mapping out of different elements, enabling us to make detailed images.[19]

There is good reason to think that some birds exploit similar physics to sense the geomagnetic field by means of *cryptochromes*, biological devices that can act as magnetic sensors.[20] Cryptochromes are proteins composed of more than 4,400 atoms arranged into various sub-molecules. Of these, two are of particular interest: tryptophan (TrpH), an amino acid related to the neurotransmitter serotonin; and flavin adenine dinucleotide (FAD), a ubiquitous enzyme first extracted from cow's milk. These two molecules, which lie adjacent to one another at a specific locale inside the larger cryptochrome structure, together enable some novel physics. Specifically, there is a particular pair of electrons in the TrpH molecule that can be readily split up, as the nearby FAD molecule features an inviting location for one electron to reside. Should an electron hop from the TrpH to the FAD, leaving its partner behind, there will now exist a 'radical pair': two spatially isolated electrons that were once coupled, but are now freed from the constraint of staying anti-parallel to each other.

Both electrons will precess independently, in accordance with whatever magnetic field they each experience. And because they are now tied to different molecules, these 'local' fields—determined by the various nearby nuclei comprising TrpH and FAD—will differ. And this means that the two electrons will wobble at different rates.

As each electron precesses, the two will sometimes be anti-parallel, sometimes parallel, sometimes something in-between. How frequently these configurations ebb and flow depends on their respective wobbling rates. Figure 4.5 illustrates how an orbital pair (a) can split into a radical pair, as shown in (b)–(d). Each electron experiences a different local field (B_1 and B_2) and the relative orientations of the spins change over time.

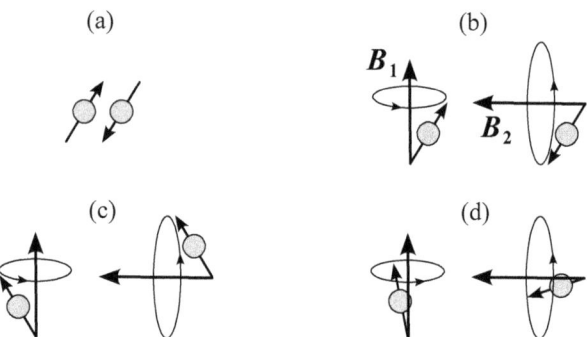

Figure 4.5 (a) An orbital pair of electrons with anti-parallel spins. (b) Separation into a radical pair, with each electron encountering different local magnetic fields, B_1 and B_2. Each electron precesses about its local field direction. The situation in (b) captures a moment when the spins are anti-parallel. (c) A later moment when the precession has brought the electrons to a parallel orientation. (d) Another configuration as the pair evolves, with the spins neither parallel nor anti-parallel.

A useful analogy to keep in mind is that of two cars waiting in a left turn lane with their signals blinking. At some particular time, both cars are flashing in sync, but later they'll be out of step, and later still, back in phase. This is an inevitable result whenever two oscillation rates differ, and how often they find themselves synced up depends on how different the two rates are.

When the electrons are anti-parallel—like the blinkers being in sync—the wayward electron has the option to hop back with its partner. This back-reaction is permissible but not mandatory. In the atomic world, such events are ultimately random, but they do accord with probabilities that depend on the local details; these are probabilities that we can calculate. Just as importantly, we can be certain that during periods when the spins are not anti-parallel, the back-reaction is forbidden; the electron field cannot support such a thing. Like two jigsaw-puzzle pieces that have been pulled apart and are now turning independently, the spins can only snap back into place on those occasions when they line up.

To review, we have an orbital pair of electrons within a protein that can split into a radical pair, which might then revert back to its initial state, but only during particular time intervals. The time intervals depend on the magnetic fields each electron experiences. To exploit this and make a biological compass, we'll need some additional functionality.

Here is another crucial bit of chemistry: after an electron hops from the TrpH to the FAD, it has produced an unbalanced charge situation, such that a proton from the TrpH molecule has some incentive to move over to the FAD as well. This 'deprotonation' may or may not occur (again, we can only speak in probabilities), but it won't depend on the parallel or anti-parallel state of the radical pair. And critically, a deprotonation event will have a chain reaction through the entire cryptochrome, recognizably altering it, so that an external feature of the protein will inform us if deprotonation occurred. In other words, there will be an external 'signal' that belies the internal states of the protein.

Let's now go back to the precessions of the radical pair and consider several scenarios. Suppose that the two electrons happen to experience comparable fields, so that they wobble at nearly the same rate. In our turn-signal analogy, we will have long spells in which everything is in sync, providing ample opportunity for the pair to revert back to their original state before deprotonation can occur. But if they are bathed in very different fields and their rates differ, they will experience less time 'blinking' in tandem and more time forced to remain a radical pair. The latter case leaves more opportunity for the proton to hop and the protein to signal. If we have many of these proteins, then the fraction of them in a signaling state is directly tied to the difference in the precessional rates, which depend on the magnetic fields seen by each electron. We know these fields depend upon the local molecular structures, and if that is all there is, then the probability of a signaling state would always be the same. It wouldn't depend on the specific orientation of the protein in space, for example. Using the language of Chapter 1, the signaling rate would be invariant to rotation.

But the internal magnetic fields produced by nuclei are not all that matter. An unpaired electron will precess based on the *total* magnetic field it feels, and there is an 'external' contribution that we've neglected so far: the magnetic field of the earth. Magnetic fields are additive, and the presence of a field external to the protein can have momentous effects. This is best seen by sketching out a few cases. Figure 4.6 shows local fields inside a protein both without (a) and with (b) an external magnetic field. In (a), rotating the protein cannot affect the internal fields relative to one another. In (b), a global field is present, and we

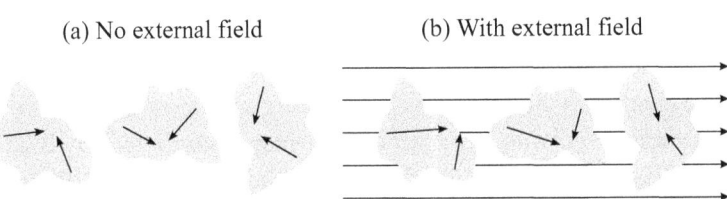

Figure 4.6 (a) Schematic of a protein and the internal local fields affecting the radical pair, shown in three orientations. The internal fields maintain the same relative configuration. (b) An external magnetic field is applied, which breaks the symmetry. The protein orientations are the same, but the net internal fields are 'pulled' to the right. The field (and hence precession) for each electron in a radical pair will now depend on the orientation of the protein relative to the external field.

must calculate the total field by adding the internal and external contribution. Fields are vectors—having direction as well as strength—so the addition can be readily performed by summing the arrows that represent the fields. Note how each arrow in (a) is, in (b), pulled to the right by the external field. *The external field breaks the symmetry.* The precessional details, and hence the probability of a signaling state, will depend on the orientation of the protein relative to the earth's field.

We now have the basis for a chemical compass, but we yet need a bit more engineering. We can exploit all this by using a systematic array of proteins with a variety of orientations, and a way to monitor their signal states. One way would be to anchor them to a spherical surface, giving us a matrix of proteins laid out with all possible orientations and a convenient common back-plane on which we can connect 'wiring' to detect each protein's signaling state.

Such a site comes ready-made with most animals, birds included. If we had our proteins properly attached to the retinal wall of an eye, it would tick the boxes we need. Moreover, eyes provide a key ingredient that our sensors will require: the energy needed to kick-start the process. Recall that we must first break up a stable pair of electrons, and we never discussed how this was to be done. The TrpH-FAD radical pair, it turns out, can be created upon exposure to blue light. Do we find cryptochrome in bird retinas? Indeed we do, and some migrants tellingly produce higher levels of it during spring and fall migration seasons.[21] What's more, experiments show that birds require exposure to daylight to sense magnetic fields.[22] We would also expect that the presence of RF fields could scramble this sensory mechanism by altering the precession and reaction details; and this has also been seen repeatedly in experiments.[23] Moreover, another optical effect—the polarization of skylight—has been shown to play a role in bird navigation, and is consistent with this mechanism.[24] We will pick up on this in Chapter 7. So, while some minutiae remain unclear, the case for the cryptochrome compass is compelling.

To recap: blue light can split a certain pair of electrons within a cryptochrome molecule, creating a radical pair. For as long as the radical pair persists, another event—involving deprotonation—may occur, which will alter the protein's shape, causing a signaling event. But the radical pair might first relax back to the original state. The probability of either outcome depends on the cryptochrome's orientation in an external magnetic field, such as that of the earth. An array of these proteins pointing in various directions makes a novel magnetic sensor, and various lines of evidence suggest that some birds have such a device built into their eyes to help them navigate.

Avian magneto-reception is sometimes breathlessly framed as a kind of 'real-world' manifestation of mysterious quantum behavior. It is indeed a quantum effect, and a novel one, but we should recall that *all* chemistry flows inexorably from quantum rules. We could just as easily point to the behavior of water, the physiology of vision or myriad other examples in which non-intuitive quantum phenomena are behind the features of our everyday experiences. We'll encounter it again when we consider the origins of sunlight and color in later chapters.

Chapter 5

By a Forest Pond: Impacts, Waves and Sounds

Imagine a visit to a forest pond, a locale rich with bird sounds, from woodpecker percussion to the ghostly notes of hidden thrushes. A kingfisher drops, breaks the water's surface, and erupts back upward, leaving concentric bullseye rings to expand and fade. In this chapter, we'll connect these seemingly disparate things: from the plunges of diving birds to hammering bills, to waves and sound, to the production of birdsong in the avian syrinx.

Impact and drag

A plunging kingfisher forces us to reckon with the three familiar states of matter: solid, liquid and gas. There's a rock–paper–scissors flavor to it: the (essentially) solid bird slipping through the same air that a moment ago held it aloft; the water, far more compliant than a solid but some 800 times more dense than the air; the liquid easily flowing, yet capable of deforming rigid objects upon impact. These states—palpably distinct and immediate to us—differ, not because of what their constituent parts *are*, but only in how they are assembled.

Air is easily overlooked, yet a body cannot pass through it without being affected. Walking presents no noticeable impediment because air hampers our motion by an amount dependent on its density and our speed. Both are low, so air resistance is often imperceptible.

If a kingfisher were to drop 5 meters, it would strike with a velocity of about 10 m/s (22 mph), some seven times faster than we typically walk. All else being equal, that means a fifty-fold increase in air resistance. But all else is *not* equal. The bird's smaller size and streamlined shape also matter. An expression for the magnitude of this resistance—or *drag*—D, which acts against the direction of motion, captures these effects:[1]

$$D = \frac{1}{2}\rho C v^2 A$$

Here, ρ is the density of the fluid, and A is the cross-sectional area the body presents as it moves at a velocity v. C is the drag coefficient that depends upon the object's shape and reflects how 'streamlined' it is. This equation applies for liquids also; so in moving from air into water, there will be a dramatic increase in drag because the density abruptly increases. As the kingfisher breaks the surface, the larger drag force will decelerate the immersed portions—starting with the bill and head—while the rest of the body continues down at full speed. This causes potentially damaging *compression* to the bird's upper body. Minimizing the impact force is therefore important, not just for kingfishers but for other plunge divers, such as gannets, which have been observed striking the water at speeds up to 24 m/s (54 mph).[2] Because the cross-sectional contact area can be reduced only so far—a bird's cross-section can be made only so small—nature must minimize the drag coefficient, which is done by tuning the bill's shape.

Drag occurs because moving molecules out of the way requires work; we must exert a force through some distance, and the reaction force impedes us. Air drag is often low because the molecules are far apart and easily moved. But liquids are dense and resist our efforts to compress them. Large forces are required to move their molecules, and our best recourse is *leverage*. Shapes that cut through fluids—such as the front of a speedboat—are tapered like an ax-blade because a wedge profile reduces the applied force needed to compress the material.

A wedge is an inclined plane, a device often employed to reduce the effort needed to raise heavy objects. Like other simple machines, such as levers, pulleys and screws, it allows us to swap distance for force, so to speak. The work we must do is the force multiplied by the distance; it doesn't matter how we allocate them, as long as their product remains the same. Make an inclined plane half as steep, and we need half the force to move an object up it, at the cost of having to move it twice as far.

For an ax, the downward force we apply is re-directed sideways and is multiplied to the degree that the length of the blade exceeds its width, enabling us not only to break the molecular bonds of the hard wood, but also to compress the material to the side. When the kingfisher's bill plunges into the water, it needn't break bonds, but it must yet squeeze the nearly incompressible liquid sideways, which comes with a huge reaction force. A highly tapered bill reduces how much of that force is directed back up in the vertical direction, minimizing the drag and compression of the upper body. No less important is that, with less deceleration upon impact, the bird more rapidly crosses the space between the surface and its nimble prey.

Despite their similar appearance, not all kingfishers are piscivorous; many species forage terrestrially and never dive for fish. A recent study asked if the diving kingfisher bills were better shaped to reduce drag upon impact.[3] Accurate replicas of bills and heads for 31 species—including both divers and non-divers—were used to estimate the impact forces from steep plunges into water. The average peak deceleration for divers was found to be about 40% lower than that for non-diving kingfishers. The relative width of the bill (the side-to-side thickness at the base) was the feature that differed between groups, being over 10% narrower on the former, on average. That divers employ a more highly tapered 'ax' to slice water demonstrates that bill morphology isn't driven purely by the requirements of handling food items. This theme will return in Chapter 11 when we see how bill shape may help regulate body temperature.

To get a sense for such impacts, we can express deceleration in terms of the earth's gravity; that is, we ask *how many gs* were experienced (where 1 g is equal to the acceleration due to earth's gravity). Measurements using models of plunge-divers indicate something in the range of 5 g for a 12 m/s speed at impact (perhaps a fast kingfisher), increasing to values between 10 g and 40 g for birds striking the water at 23 m/s (more typical for gannets).[4] These numbers may seem large, until we consider those experienced by members of a different bird family obliged to engage in even more violent collisions.

In the late 1970s, an Acorn Woodpecker *Melanerpes formicivorus* with an injured wing took up residence at a California park ranger's office and would drum on a trunk in response to the clacking of typewriter keys.[5] It became the subject of high-speed photography that was analyzed to reveal that its hammering involved an average deceleration of around 700 g, with some impacts generating over twice that value. Humans suffer concussions following deceleration in the range of 135 g, so understanding how woodpeckers tolerate such abuse is of considerable interest.[6]

A popular explanation—that spongy bone in the bird's brain-case acts as a 'shock absorber'—sounds plausible but isn't correct. If force were dissipated between the bill and the skull, it would also reduce the penetration into the wood, making the motion less effective. The bird would simply have to hit harder.[7] This is borne out by a recent study of three woodpecker species showing that the brain undergoes deceleration comparable to that of the bill upon impact.[8]

How is this possible? Simply put, deceleration alone does not determine the risk of injury from a collision.[9] The *duration* is no less important. For humans, a 160 g impact lasting 3 milliseconds will be as tolerable as 80 g acting for

15 milliseconds. Woodpecker collisions involve much shorter time scales, on the order of 0.5 to 1 millisecond, permitting much larger deceleration. Equally important is the effect of the brain's size. A human brain experiences far more force than a woodpecker's brain at the same deceleration because it is far more massive. What will ultimately determine if tissue is damaged is the *stress σ*, which is force divided by area. Assuming that brain tissues of woodpeckers and humans tolerate comparable stress and have similar density ρ, we can equate the stresses and write them in terms of volumes V, areas A and decelerations a as follows:

$$\frac{\sigma_H}{\sigma_W} = 1 = \frac{\rho V_H a_H / A_H}{\rho V_W a_W / A_W} = \frac{r_H a_H}{r_W a_W}$$

The subscripts 'H' and 'W' refer to human and woodpecker, respectively. The r terms are the skull radii, and we assume spherical heads for simplicity. This tells us that, at about one eighth the radius of our brains, woodpecker brains can tolerate eigth times the deceleration. Moreover, differences in shape increase this by another factor of two. Combined with the effect of short duration, it is estimated that even 6,000 g impacts might be tolerable for some woodpeckers. Modest changes in length and time scales can have surprisingly large effects.

Water and waves

No matter how streamlined, a plunging kingfisher still makes a mess upon breaking the liquid barrier. As the bill first enters and leverages water molecules aside, some are forced upward, and others are ejected outward. Once submerged, the bird pulls air behind it, forming a cavity: a short-lived air pocket that invites the water to collapse back in. Like a pendulum swinging past its stable point, liquid overfills the cavity, colliding with itself as it splashes over the water column. Soon, the bird dashes back through, creating another disruption.[10]

Every eye-blink dance of water from such a plunge has a snowflake-like uniqueness. In a relative way, all splashes are the same, but their detailed decorations are not. Even dropping the same sphere from an identical height, we would be hard pressed to record two identical splashes. Yet the aftermath of such collisions always resolves into the same nested, expanding bracelets of changing water heights, nature's most familiar example of wave motion.

The importance of waves cannot be overstated. All matter and energy is ultimately built from them, and they manifest in all manner of phenomena, including many that will be topics in coming chapters. It is impossible to

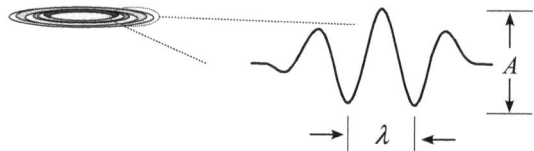

Figure 5.1 On the left is an expanding ripple on a flat water surface. On the right is an approximate depiction of the expanding wave, a cross-section of the ripple. The vertical distance from trough to peak is the amplitude A, and the horizontal distance between troughs is the wavelength λ.

understand sound, light, color or optics without a good intuition for waves, so it behooves us to dwell on them at this larger scale in some detail here.

Waves are simple yet subtle, even paradoxical. They can cover vast distances without any *thing* moving far. The pond's various water molecules momentarily go up and down, this way and that, soon returning to where they were. Their motion is *local*; the expanding rings do not carry any water away, like the current in a river would. The water at a splash site remains *there*. Only the disturbance actually travels. This is more obvious with the wave sometimes made by fans at large sporting events. With everyone nominally seated, a group in one section stands up momentarily, intending their neighbors to one side to take the cue. Soon there is a propagation of standing and sitting, resulting in a disturbance that moves sideways through the venue, although not one person ever moves sideways. People merely go up and down—this is what we mean by local movement—but the wave moves freely from one locale to the next.

Imagine that we freeze time so as to inspect the pond's wave more closely. As shown in Figure 5.1, we can take a cross-sectional view of the ripple, a profile of the water's shape. We'll find the pond's level here is modulated by few crests separated by troughs, falling away on the sides. The underlying shape reflects the ubiquitous sinusoid that we found in Chapter 1 when we traced out the sun's height.[11]

Two numbers characterize this snapshot: the vertical displacement from trough to crest, the *amplitude A*; and the sideways distance from crest to crest, the *wavelength* λ. At this single point in time, we see the spatial profile of the wave. But we could have looked at one point in space and monitored its time dependence. From the perspective of an insect resting on the water's surface, the oncoming wave approaches at some velocity, v, and as it arrives, both water and insect will move vertically—completing several cycles up, down and back—each taking a span of time, the *period T*. We could also speak of $1/T$, the *frequency f* of motion, which measures the number of cycles per second. If we recorded the height of the insect, and plotted it versus time, we'd get a picture identical to that in Figure 5.1, with period T replacing wavelength λ.

These numbers tie up nicely in a way that isn't hard to see. One wavelength advances as one period goes by, so the velocity—the distance divided by time—has a magnitude given by:

$$v = \frac{\lambda}{T} = \lambda f$$

Velocity, of course, entails a direction. For the insect, the wave moves from the source directly toward it. But a second insect on the other side of the pond will report a different direction. Stepping back, we see that a single arrow cannot account for the velocity, as it does for the kingfisher's falling motion. Instead, the wave moves outward in *every* direction; an infinity of arrows are needed. We might represent this with just a few of them; pointing north, south, east and west—and the directions in between—making a wheel with eight spokes.

The Latin word for wheel is '*rota*'; hence terms such as *rotate*, *rotational*, *rotor* and *rotary* are all concerned with a turning motion. The word for spoke is *radi*, leading to words such as *radius*, the measure from the center to the edge of a circle, and *radiation*, a term saddled with the negative connotations of a nuclear disaster. But radiation means something much more general: to *radiate* is to do exactly what our water wave is doing, to simply move out, away from a source in various directions, just as a *radiator* expels heat every which way, and a *radio* tower broadcasts its signal outward. Certain materials that spew out particles without any prompting are *radioactive*. And light, which also radiates from a source, can be broken into slivers—say, along one of those eight spokes—so we call that a *ray*.

Wave behavior becomes rich with possibilities when there is more than one. When multiple waves are present, they combine in a specific way to make something new—we speak of this as the waves *adding*. To illustrate what this means, we can start with two wave sources. We'll expand on a model we introduced with electromagnetic waves in Chapter 2. Suppose that two fishing rigs, consisting of bait under bobbers, were cast out on our pond and have been sitting motionless. Several fish arrive and begin teasing their meals off the hooks (let's imagine, in perfect synchronization), setting both bobbers into gentle undulating motion.

Figure 5.2 demonstrates how the waves caused by this motion will combine. Specifically, we'll consider the point midway between the two bobbers, as shown in (a). At the moment corresponding to (b), the waves have not yet met. We show each individual wave as dashed lines, and the overall water height as a solid line. In (c) the leading crests occupy the same place, and additive effect produces a high peak; this is *constructive interference*. Note how the amplitudes simply add together. A quarter period later (d) the crests and troughs line up,

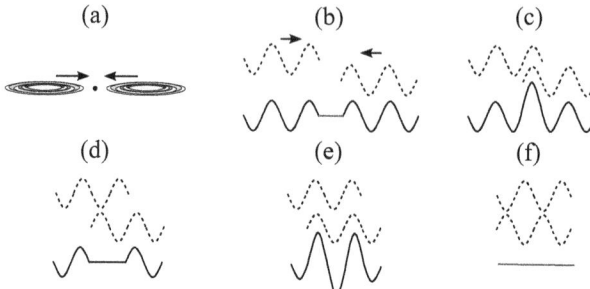

Figure 5.2 (a) Two identical ripples spread on a flat water surface. We focus on the point midway between them. (b) Dashed lines show individual waves moving in opposite directions just before meeting. The sold line shows the overall profile of the water's height. (c) Later, when half a wavelength overlaps, leading to constructive interference. (d) Now a full wavelength overlaps, but with the waves being out-of-phase, destructive interference results in zero amplitude. (e) The waves have moved on so that 1.5 wavelengths overlap and constructive interference results. (f) Even later, as two full waves are out-of-phase, momentarily leaving a large section of flat water.

so the addition corresponds to a perfect cancellation: *destructive interference*. Diagrams (e) and (f) show the continued evolution as the waves slide by, and we simply add the heights at each point.

This *superposition* of waves and the resulting constructive or destructive interference isn't restricted to the location between the two sources; it manifests everywhere, as demonstrated by another illustration. Figure 5.3 shows the scene, viewed from above, with the sources near one another at the far left, and the wave crests (solid lines) and troughs (dashed lines) moving outward toward positions marked *A* and *B*. At *A*, the sources are equally distant and the two waves always arrive *in-phase*; hence, constructive interference results. At *B*, however, the distance from one source is $\lambda/2$ further than the other. Now, when one source gives a crest, the other is a trough, and vice versa. The

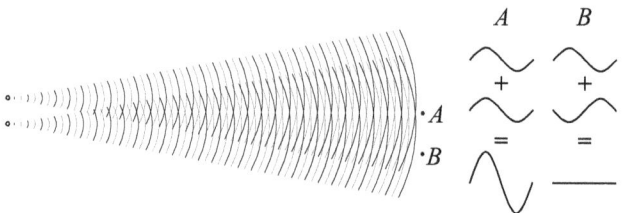

Figure 5.3 Looking down on two identical side-by-side wave sources at the far left. A portion of the wavefronts from each are shown propagating to the right. Crests and troughs are indicated by solid and dashed lines, respectively. At point *A*, equidistant from the sources, the waves always arrive in-phase and yield constructive interference. At an off-axis point *B*, the waves always arrive out-of-phase, leading to perfect destructive interference.

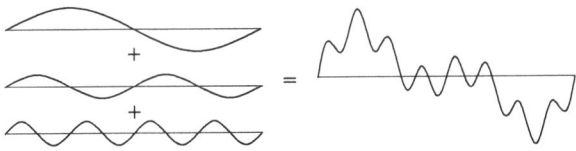

Figure 5.4 The left side shows three sine waves with different wavelengths. Adding the total amplitude at each point results in the complex waveform on the right.

waves are perfectly *out-of-phase*, and adding the results always yields zero; this is destructive interference.

These are simple examples of superposition but the principle extends to any number of waves of any frequency, amplitude or direction. In this way, complex waveforms can be built from sums of sine waves. An example is shown in Figure 5.4, where for each location along the horizontal axis, we measure the height of each sine wave on the left, add the numbers and plot them on the right. In Chapter 7, we'll see how to reverse this process, breaking a complex waveform into its component sinusoids, a procedure that is integral to making audio spectrograms.

On this note, we'll start our exploration of audio by moving from two to three dimensions, and considering a different medium for our waves, namely the air. We'll again create a disturbance based on local motions of particles, but with a new twist.

Air and sound

On a windless morning the air seems uniform and unchanging, and with a few meteorological tools we could verify as much. A thermometer would reveal steady temperature; an anemometer would record zero wind speed. Both results would accord with sensations we pick up via our skin. A barometer would report something less tactile: the atmospheric pressure, the force per unit area, which we are usually oblivious to, since the air is always pushing in every direction at once, with the forces balancing one another.

The amount of force manifested by this air pressure is not trivial, however. A credit card held aloft will experience roughly 100 pounds (445 Newtons) of force pushing against one side. But the air on the other side pushes back just as hard, negating it. So, in order to measure this seemingly hidden property of the air, we require a device that only senses force in one direction. A closed cylinder with a freely sliding piston will demonstrate the extent of air pressure if the air inside is removed. If it were the size of a soup can, about 100 pounds of force would slam the piston inward. If we instead exert a known force in the other direction—perhaps placing a spring inside that pushes back harder as it is compressed—we'll be able to measure the force and hence the pressure. No matter how we orient this crude barometer, we'll find the air pushing against the piston with the same force.

The origin of pressure can be understood by zooming in. Up close, we'll find that air is mostly empty space, bubbling with speeding molecules that occasionally collide. Consider a very small cube, something we can just barely see or imagine, with sides 0.1 millimeter long. It would contain about 10^{13} (10,000,000,000,000) molecules moving at many hundreds of miles per hour, but they'd take up so little room that 99.9% of the volume would be empty space.

Unlike the water in the pond, which is some 800 times more dense, the air is compressible and 'springy' because with all that empty space, the molecules can be forced to occupy a smaller volume if we push them into it. But they don't go without a fight; squeeze a balloon and the gas inside pushing back is palpable. Moreover, the molecules we compress will push on those beyond them, and like a chain of dominoes going every which way, the effects of our action will spread out. This should start us thinking about wave motion.

The ripples on the pond involved vertical displacement of the water level, while the disturbance moved horizontally. It is a similar situation for the fans in a sports venue, themselves standing and sitting while the wave moved sideways. These are examples of *transverse* waves: the motion is at right angles to the direction in which the disturbance travels. What happens when we push on gas molecules would be like the fans taking a different tack—rather than standing up, they'd sway sideways in one direction. Imagine everyone in a section suddenly leaning into their neighbor to the left, pushing them leftward before they recoil to the right. This would nucleate a *longitudinal* wave.

Pushing on the air is bit more complicated. When a bee's wing, for example, sweeps forward, the frenzy of collisions in that suddenly compressed air will redistribute the force outward. The pressure change will radiate away in all three dimensions, as an expanding 'shell' of higher pressure. On the heels of this, a shell of lower pressure must follow. With the next beat of the wing, another disturbance of compressed and rarefied air is produced and sent out.

We've just described the production and radiation of *sound*: periodic changes to the local air density and pressure producing a longitudinal wave, which jostles molecules back and forth along the direction that the wave travels. Figure 5.5 shows the correspondence of a longitudinal wave, in terms of the density of molecules, and the measured density or pressure along one direction in space. Such a plot allows us to *represent* sound with a transverse wave shape.

Every disturbance must travel at some speed, and sound in the atmosphere typically moves at about 767 mph, or 343 m/s. The sounds that we are concerned with—the ones we can hear—have frequencies from about 20 Hz (cycles per

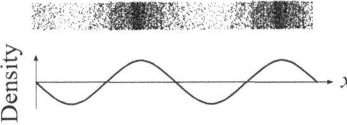

Figure 5.5 The top depicts a volume of air with a sinusoidal distribution in the density of its constituent particles. This amounts to a series of condensations and rarefactions along its length. The density, when plotted as a function of position, is shown below.

second) to 20,000 Hz relating to the pitch we perceive: lower frequencies being bass and higher being treble. In terms of wavelengths, these measure from about 17 meters down to 1.7 centimeters.

A bigger amplitude corresponds to greater displacement of molecules, which means larger changes in pressure and density and increased loudness. These changes determine the sound's *power*—the rate at which it conveys energy—which we can use to help quantify loudness. Every source delivers sound at some power, which then spreads out through space as the wave propagates. The loudness that we as listeners detect will depend on how much of that spread-out power we can collect. Since the sound radiates outward in an ever-growing sphere, the power is spread across an ever-growing area. This leads to the concept of *intensity*, defined as the power divided by area. At a distance r from a source, the power is spread on a spherical surface of area $4\pi r^2$, so for a power P, the formula for intensity (I) is:

$$I = \frac{P}{4\pi r^2}$$

We note that this is another 'inverse square' rule, as we had with gravity and electrostatic force. Nature is being economical again.

There exists a wide range of intensities that we hear—the loudest sounds we can experience without pain, perhaps from a concert or nearby thunder, have an intensity about one *trillion* times larger than the subtle sounds at our threshold of hearing. Using a scale of one to a trillion is unwieldy, so we employ a compressed scale: the decibel, or 'dB'. The minimum perceptible sound is 0 dB, while the thunder comes in around 120 dB.[12]

Finally, there is also the *quality* of the sound—the timbre—which we must account for. Timbre lets us differentiate between instruments playing the same note—an 'A' is a wave at 440 Hz, but a piano and a guitar producing this same note are easily distinguished, as would be the pure whistles of orioles, the nasal beeps of nuthatches and the discordant screeches of jays, even were they all at the same pitch. It is the *shape* of a waveform that determines timbre.

In Figure 5.4, we can see how several sine waves with different wavelengths add together to produce a more complicated wave. Provided that each wave has a frequency that is an integer multiple of the lowest frequency f_0, the resulting shape will also have that lowest, or fundamental frequency. There is an endless variety of ways that the higher frequency waves, the *harmonics*, can be added, and this is why there is such a diversity of possible timbres. To produce a complex sound waveform, then, an audio source causes the surrounding air to vibrate at various equally spaced frequencies (such as f_0, $2f_0$, $3f_0$, etc.) simultaneously. But it must also be able to rapidly change f_0 (and perhaps the harmonic content) for it to produce anything besides a simple flat tone; that is, for it to be a musical *instrument*. A bird's syrinx is such an instrument, and in the following section, we will see how it operates.

Syrinx and song

Many birds can be emulated by whistling—a process performed entirely by the mouth—so it's easy to regard the vocalization of a bird with its head back and bill open as something that its mouth is doing. Instead, its voice originates deep in the breast, in the syrinx, which is a device similar to our larynx.

Both organs exploit vibrating membranes to initiate pressure changes and modify the resulting sound waves via the windpipe and mouth before they are released to the world. For humans, the membranes are the vocal cords, which alone produce only a limited range of sounds; it is by shaping the tongue and lips and teeth that we make diverse vocalizations. With birds, however, most of the variety originates in the syrinx.[13]

Three varieties of syrinx are recognized, but we will focus on the one used by most passerines: the *tracheobronchial syrinx*, shown in a simplified version in Figure 5.6. At the Y-shaped junction lie pairs of soft membranes, the lateral and medial tympaniform membranes, or labia; these are elastic tissues that can extend into the lumen and alter the air flow between the lungs and the trachea.

These membranes are anchored to rigid cartilaginous tracheal rings, which provide support. Various muscles control the spacing of these rings, affecting the membrane tension and extension into the lumen. The membranes can be adducted inward to meet and close off the airway, or pulled tighter to the walls, opening the passage. Each side has its own enervation and can be controlled independently. Many birds utilize just one pair of membranes to produce sound, but species with especially rich songs, such as thrushes, may employ both simultaneously, like a pianist using two hands to play two different melodies.

Sound is initiated as exhaled air flows between these membranes. The air molecules, which move every which way at many hundreds of mph even in 'still' air, now acquire greater velocity parallel to the walls of the bronchial

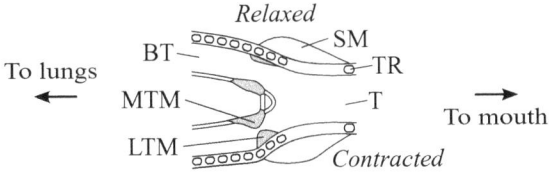

Figure 5.6 Schematic diagram of a typical passerine syrinx. Two bronchial tubes (BT) on the left lead to the lungs. On the right, the trachea (T) leads to the mouth. Tracheal rings (TR) support the structure, which can be altered in shape by the syringeal muscles (SM). Here, the top SM is relaxed, while the bottom one is contracted. The state of the SM alters the shapes of the lateral tympaniform membrane (LTM) and medial tympaniform membrane (MTM). The passage of air through the ensuing constriction causes these membranes to vibrate and produce sound. [Modified from Riede and Goller (2010).[14]]

tubes, and so they will carry less momentum into the walls during collisions. That means less force and, hence, less pressure. This is *Bernoulli's principle*; it is a consequence of the conservation of energy in a fluid.[15]

The lower pressure draws the pliable membranes inward, which further constricts the flow, causing air molecules to pile up on the bronchial side. The more the membranes are stretched, the greater their own tension, which acts as a restoring force to pull them back into place, allowing air to pass. As that air accelerates through, the pressure drops again, and the cycle repeats, setting up a combined 'flapping and rolling' motion in the membranes. The frequency of this motion—which will dictate the pitch of the sound—is largely determined by the tension in these tissues, which is controlled by the syringeal muscles. Plucking a rubber band while stretched to different extents demonstrates how easily the pitch is changed through such a mechanism.

Several musical instruments initiate sound in a comparable way.[16] With an oboe or bassoon, air is pushed between two closely spaced, flexible surfaces: the reeds, which are pulled together by the lower pressure of moving air. The reeds deform until their own restoring force pulls them back outward, re-opening the passage, completing a cycle. Similarly, lips pressed against the mouthpiece of a trumpet or trombone rapidly vibrate as air is forced through the narrow space between them.[17] Blowing through tightly pursed lips into a fist, as if warming the hand on a cold day, can replicate this effect.

The reeds of an oboe or the lips of a trombonist only nucleate the vibrations, producing weak and unmusical sounds. It is within the body of the instrument that a rich timbre and substantial volume are generated. This is due to a mass of air, largely constrained in some kind of cavity adjacent to the vibrating source. This often involves a tube that is open at one end, and the avian instrument employs the trachea and mouth to this effect. Because the cavity

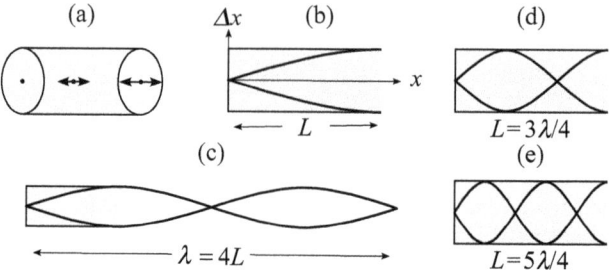

Figure 5.7 (a) A cylindrical tube, closed on the left side and open on the right. Air molecules shown as dots cannot move horizontally at the closed end, but are free to do so at the open side, as indicated by the arrows. In the middle there is an intermediate range of motion. (b) The range of motion Δx plotted versus position x along length L of tube. (c) This motion is seen to comprise a quarter of a full wavelength, which extends beyond the tube. This condition is the longest wavelength, lowest frequency mode. (d) The next shortest wavelength, having maximum amplitude at the open end. (e) The next allowed configuration, allowing us to see the trend is $L = N\lambda/4$, where $N = 1,3,5$, etc.

not only amplifies the sound but modifies its timbre or quality, the process is sometimes called *filtering*.

Sound radiating through the open air is simple to describe. When it is largely constrained within a chamber, however, the motion is more complex, because there are boundaries that cause waves to reflect and the reflected waves superpose together. This can get complicated, so we'll focus on the simplest cases to develop a sense for how an air cavity affects sound.

Figure 5.7 (a) shows a cylindrical tube, closed on the left side and open on the right. Air inside feels the same pressure as the surrounding atmosphere, but it is not as free to move. This is indicated with the arrows. At the open end, air molecules have a full range of motion, but at the closed end they hit a wall. The range of motion will vary between these extremes; in the center, there is an intermediate level of motion possible. In (b) we plot this range of motion Δx versus position x for the tube with length L. In diagram (c), we can see that the variation of motion in the tube of length L is one quarter of the wave that extends outside the tube. In (d) and (e), we see that other modes with shorter wavelengths are possible. In every case we have zero amplitude at the left end—a point is referred to as a *node*—and maximum amplitude on the right end, forming an *anti-node*.

It is not difficult to see that the tube length must correspond to an odd multiple of a quarter-wavelength. Hence $L = N\lambda/4$, where $N = 1,3,5$, etc. Because wavelength depends on velocity and frequency according to $\lambda = v/f$, we can rewrite this as $f = Nv/4L$. The lowest frequency naturally supported in the tube is for $N = 1$, the fundamental frequency, f_0. All others are odd integer multiples.

A tube that is open at both ends, or that has a sound source at one end as opposed to a stationary wall, will support a different set of frequencies. The critical takeaway for us is that every cavity has a set of *resonant* frequencies, or modes, determined by its shape. When a sound source—even of only a single frequency and low amplitude—is placed near such a cavity, it will cause various modes to be excited, producing both amplification and filtering.

The syrinx can thus be modeled as a *source*, the vibrating membranes that initiate the sound, plus a *filter*, the air column of the trachea and mouth, which amplifies and further 'shapes' the signal by accentuating its own natural frequencies and attenuating others. This provides a number of ways to fine-tune the sound (opening the mouth to varying degrees, for instance, to change the filtering effects of the air column).

To conclude, we'll focus on the two key controls that allow each song syllable to be constructed: the pressure P of the air driven through the syrinx, and the 'spring-constant' k, which controls the stiffness (and hence the vibrational frequencies) of the membranes.

The formation of different syllables is determined by how k and P change over time. Stated more abstractly, *each syllable is defined by some path through the k-P parameter space*. This may seem an arcane way to phrase it, but Figure 5.8

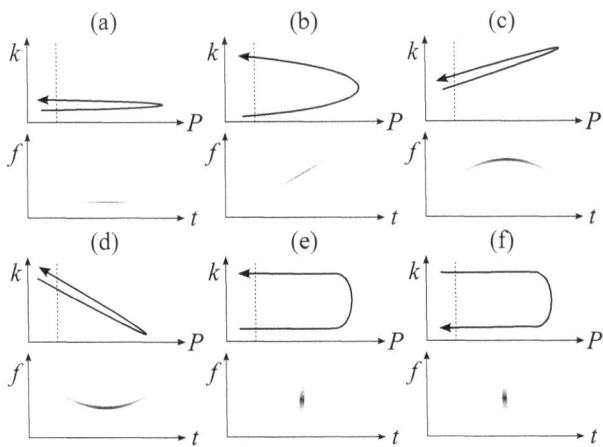

Figure 5.8 Example changes in k–P space and the corresponding syllables produced. In each of the six cases, the top plot shows a path through the space of k (spring constant) and P (pressure) values. The dashed vertical line indicates the minimum pressure needed to cause vibration. The lower plot for each case is the corresponding spectrogram showing frequency f versus time t, with darker gray corresponding to louder sound. (a) Increasing then decreasing P at fixed k yields a single tone. (b) Increasing then decreasing P as k goes up yields a simple rising pitch. (c) Increasing P and k together before decreasing them in tandem yields a rising and falling note. (d) Increasing P while decreasing k, and then reversing the process yields the reverse case. (e) A counter-clockwise path through the space producing a sudden up-slurred note. (f) A clockwise path producing a similar down-slurred note. [Modified from Mindlin and Laje (2005).]

illustrates the utility of framing it this way. Each diagram (a) through (f) shows how the manner in which k and P are varied (top) gives rise to different basic shapes seen in spectrograms that plot frequency versus time (bottom). The dashed vertical line represents the pressure threshold needed to initiate sound. The solid line (the 'path') in each k-P diagram is simply the instruction for how the two parameters are to be varied over time.

This model is simple, but it has considerable explanatory power. Coordinating breath and syringeal muscle tension traces a route through k-P space, producing an elemental note or slur that can be strung with many more to form a song. Methods for generating even greater acoustic variety are not hard to envision. Using both sides of the syrinx independently is an obvious way to add complexity, but is not the only approach.[18] Various chickadees, for example, employ the two sides in tandem, coupled with a kind of feedback effect between them; the 'dee' notes of their eponymous call specifically result from such an interaction.[19]

In other cases, the filtering effect of the trachea may affect the operation of the source, and such feedback effects cannot be captured with our simple model. Other birds may use a different filter altogether. Doves inflate a portion of the esophagus, adjacent to the syrinx, to create a sounding body that generates volume without the bird needing to open its bill.[20]

Having a sense for how birds vocalize and how the signal propagates through air, we'll turn next to the physics of sensing sound. Audio detection has a number of important applications, including: communication; detection of prey; and even echolocation, a tool surprisingly used by several bird species.[21] We'll also explore the elegant concepts that allow us to record and analyze audio in the following chapter.

Chapter 6

Under Night's Cover: Hearing, Recording and Analyzing Birdsong

Birding is no less about sound than sight, and many of us first attune our ears while under the cover of darkness. In this chapter, we'll consider our own aural capabilities and the more sensitive hearing of owls. Because the recording of bird vocalization is so integral to ornithology—and is a growing hobby within birding itself—we'll explore the associated technology of microphones and digital recording, and look at the mathematics behind the ubiquitous tool of sound visualization: the spectrogram.

Owls and ears

Birding by ear might seem an afterthought, something to be learned after our visual skills have been honed or when we need a new challenge. The comment 'heard only' made on a checklist carries a certain forlorn tone, a tacit disappointment. This mindset does us a disservice. Yes, we'd like the best possible view, but why does the reverse rarely generate a complaint? If on an international trip we managed to merely see, but not hear, a Bearded Bellbird *Procnias averano*, or a Capuchinbird *Perissocephalus tricolor*, or an Eastern Whipbird *Psophodes olivaceus*, wouldn't the event be deserving of a regretful 'seen only' field note?[1]

We may be visual creatures, but in many ways an auditory connection with a bird is far more profound. Unlike a mere reflection of the ambient light, every bird song is an idea that an avian brain has learned, practiced and produced by an ingenious respiratory system, freely offered up for our edification. Our own neurons fire in synchronicity, in a kind of avian-to-human mind meld that we have the privilege of experiencing and, to some degree, understanding. And the palette is objectively far broader: our hearing spans a thousand-fold range of frequencies; our vision spans a mere factor of two.[2]

Hearing can be thought of as the final step in a communication channel, in which a source encodes a signal and broadcasts it to one or more receivers. But a receiver must do more than detect the signal. It must faithfully capture how it changes. In the case of sound, it must register changes in volume, pitch, timbre and so on; and it must update our conscious experience accordingly.

A receiver is a type of *transducer*: a device that changes the type of energy used to carry a signal. Since brains function on the basis of electrochemical signals, a sound transducer is needed to convert acoustical to electrical energy, preferably with a high level of fidelity, keeping the signal's content preserved.

Most avian vocalization, it would seem, serves the purpose of within-species communication. Luckily, our range of hearing overlaps theirs, permitting us to eavesdrop. Our ear physiology is also rather similar to theirs, with a few exceptions, such as the external outer-ear structures that most mammals share, but birds lack.[3]

Both birds and humans have ear canals—flared openings that funnel into an air-filled cylinder, terminating at a vibrating membrane, the eardrum. That these structures resemble a wind instrument (or vocal tract), as described in the previous chapter, is no coincidence. The air within an ear canal—like that in a woodwind instrument—forms a resonant chamber, helping to amplify a range of frequencies.

The eardrum is like a vibrational source run in reverse: instead of oscillatory motion initiating the condensation and rarefaction of adjacent air, the flexible surface reacts to pressure changes, being pushed or pulled accordingly. This is the first job of the ear-as-transducer: to convert acoustical energy to mechanical energy.

The vibrations of an eardrum are tiny. Even loud sounds, such as city traffic, cause the membrane to move just tens of nanometers.[4] But mechanical contrivances, such as linkages and levers, are employed to scale up this motion. Humans utilize three small bones to couple the eardrum movement to the so-called oval window of the cochlea (where nerves will transport signals to the brain), amplifying the signal by a factor of 1,000 or so.[5] The avian ear does the same thing, but with only a single, straight bone, employed as a lever, which—similar to the handles of a wheelbarrow—trades movement for force.

The cochlea is a complex organ that is much more sophisticated than the membranes and bones that supply it with an amplified mechanical signal. Shaped like a long cone, it is filled with liquids that support longitudinal waves created by the vibrating oval window that lies at its base. Its name comes from the Latin for 'snail', since in mammals this cone is wrapped into a tight spiral shape. In birds it is not coiled at all, but extends straight.

The cochlea does the heavy lifting of the hearing process, transforming the signal a final time by converting longitudinal waves in the fluid into nerve signals. The process isn't trivial, and we'll necessarily gloss over many details. Down the length of the cochlea are rows of specialized hair cells that are sensitive to the movement of the surrounding fluid. Each cell has its own neural connection; so the 30,000 nerves that line the cochlea, bundled like a parallel cable, are each mapped to a specific position. This is critical, because the cochlea interior is constructed such that different wave frequencies will have larger amplitudes at different locations along its length. Near the oval window, high frequencies dominate, while the lowest frequencies propagate to the far end. This *tonotopic mapping* makes the cochlea function like an 'acoustic prism' that takes a mixture of sound frequencies and disperses them along a line of sensors.

The hair cells—as the name implies—are long projections into the surrounding, ion-rich fluid. Here, the cell membrane contains ion channels, which remain closed until they are leveraged open when the hairs are pushed to one side. The surrounding fluid contains potassium and calcium ions, which will rush in when these gates are opened, triggering the cell to release neurotransmitters and set off a nerve impulse.

Our hearing deteriorates over time, often due to damage to these cells, which is permanent. Beginning in the 1980s, investigators found that birds don't have this problem; they could replace damaged cells by generating new ones.[6] In a matter of several weeks, the reconstruction can be done so well that it is difficult to see any difference between an undamaged and a recovered organ.

Most birds hear in a narrower frequency range than mammals do, topping out around 10 kHz. However, birds are capable of better temporal resolution; that is, they can make out details in rapidly changing signals that will sound 'blurred' to us.[7] When we listen to a slowed-down recording of rapid trilling, for example, we'll often pick up subtleties that would otherwise pass too quickly for our perception—features of the song that are apparently clear to avian ears.

Some bird species—especially nocturnal ones—employ hearing for more than communication. Many owls use sound not merely to detect prey but also to effectively pinpoint its location. Some of the anatomical adaptations to this end are quite clever.

Two ears are far more useful than one because various properties of the signals they deliver, such as their time of arrival and intensity, will in general be different from one another, depending on the source location. Assuming two identical ears on either side (as is our case), a sound from directly in front reaches both simultaneously and with the same power. If the source moves to

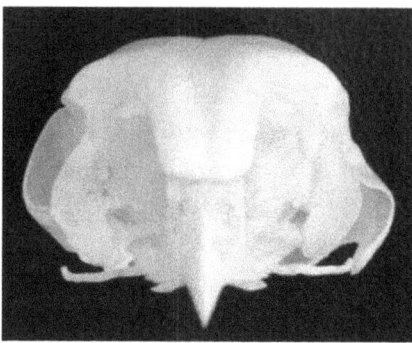

Figure 6.1 The skull of a Northern Saw-whet Owl *Aegolius acadicus*. The asymmetry in the location of the ear cavities is apparent. [Photo credit: author.]

one side, the sound arrives at the closer ear first, which may be detectable and be used to provide a sense of direction.

If a sound source moves directly up or down, however, it poses a greater challenge because each ear will receive the same signal. And if the ears share the same spatial sensitivity pattern, there will be no way to deduce the source direction. Various owl species have solved this problem by breaking the bilateral symmetry.[8] The right ear on a Northern Saw-whet Owl *Aegolius acadicus* resides significantly higher on the head and is canted upward, while the left sits lower and is directed downward. Because each ear has different sensitivities to the vertical placement of a source, comparing their output signals allows for vertical discernment. The striking asymmetry is hidden by the feathers, but is evident from a glance at the skull, shown in Figure 6.1.

Another acoustic adaptation is one of the most immediately recognized features on some owls: the facial disk.[9] The concave shape of an owl's face helps collect sound waves like a satellite dish, partially reflecting waves inward to the ears, instead of back outward as a convex head shape would. More on this when we discuss parabolic microphones, below.

We can let our ears be the final link in the chain, enabling us to simply enjoy bird song and leaving it at that. But vocalizations are an integral part of the avian phenotype, no less valuable for study than the bodies and nests curated in museums. To capture and preserve the transient messages that birds broadcast requires that we fashion clever tools that are just as interesting as those that nature has evolved. This is something new to us; we've only had sound recording technology for some 150 years, less time than we've had photography.[10] The remainder of this chapter explores how we collect and analyze birdsong, beginning with microphones and ending with spectrograms. The material is necessarily more technical than any so far.

Microphones and their minutiae

The experience of hearing—having the sounds of nature, or language or music, play out in our conscious minds—is uncanny to contemplate. Imagine attempting to describe the timbre of a violin vividly enough for a deaf person to reconstruct it. It isn't possible, for nothing can stand in for the perceptual experience of audio, any more than it can for light and color. Despite this, we can yet capture and quantify sound. The technologies for doing so have become so ubiquitous that we hardly find it surprising, until we realize one day that our phones are identifying bird songs.

Prior to the mid-1800s, the only 'recording system' for (some) sounds was sheet music, which obviously was not amenable for capturing much else.[11] Lacking the sophisticated electrical circuitry that we are accustomed to nowadays, early field recording technology was 'low-fi' and bulky. This did not prevent workers in the 1930s from doing field recordings; an early effort involved a mule-drawn wagon being used to transport a hefty tube-based amplifier and recording device out to record Ivory-billed Woodpeckers *Campephilus principalis*.[12] Over the next decades, professionals could take cumbersome equipment into the field to capture birdsong, but portable, affordable options only became available in the 1960s, with the launch of the cassette tape format.[13]

The front-end of a recording system is the microphone, which, like an ear, transduces audio into a usable electrical signal. Their varieties are endless, as are their specifications, merits, liabilities and nominal applications.[14] They typically utilize a diaphragm that plays the role of the eardrum, converting pressure waves into mechanical motion. This then drives some kind of *element*, for which positional movement affects its electrical properties.

For the recording of birdsong, one typically employs a *condenser* microphone, which makes use of a *capacitor*: a charge-storing device. A simplified picture of a capacitor as a microphone is shown in Figure 6.2. The capacitor here is merely two parallel surfaces ('plates'), each of which will readily hold an equal but opposite electric charge. A charged capacitor has potential energy, and hence a certain voltage, which is simply the energy per charge.

A given capacitor's voltage depends on several factors, such as the separation of the plates. Two oppositely charged plates experience an attractive electrostatic force—so it will take work to move them apart—which results in a greater potential energy and, hence, a larger voltage. Keeping one plate fixed and allowing the other to move in response to air pressure, as shown, produces a voltage that varies in lockstep with the sound impinging on it.

This is an elegant design, and the basis for a sensitive and accurate device. The capacitor, however, needs to be charged up, or *biased*, which typically

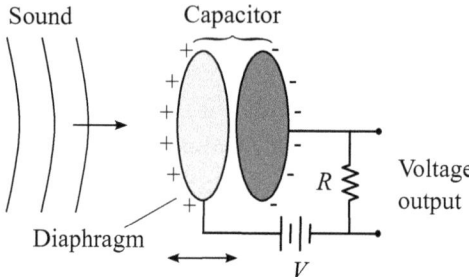

Figure 6.2 A condenser microphone transduces an acoustic signal into an electrical one. Sound waves arriving from the left cause the diaphragm to move in concert. Here, the diaphragm is one side of a parallel-plate capacitor attached to a voltage source V. The change in its position alters the amount of charge that the capacitor can hold. The resulting voltage across the resistor R varies as a result.

requires external power. Some microphones skirt this with a 'permanent' charge source to keep the plates full: namely, an *electret*, which is analogous to a magnet, but with fixed positive and negative charges populating each side instead of north and south poles.

Electret or not, the voltage fluctuations across the element are typically so small as to be useless without additional circuitry to amplify them and create a viable signal. This requires that *active* electronics (that is, electronics that must be *powered*) be incorporated, which adds complexity and cost and reduces robustness.

Many condenser microphones today are integrated into smartphones (where power is readily available), and are known as MEMS (micro-electrical-mechanical systems).[15] They are fabricated using processes developed for semiconductor electronics, and benefit from similar improvements in scale reduction. Their performance can be surprisingly good in light of low cost being a primary driver of their design, but they still fall short of devices specifically engineered to overcome obstacles to fidelity. These obstacles are broadly categorized as noise and distortion. The former is unavoidable and especially bothersome every time we amplify our signal, because in doing so we make hiss, static and other undesirables more noticeable. Distortion can be more subtle, but manifests in various ways, such as via a non-uniform frequency response that alters the timbre (or 'colors' it), or by failing to handle large signals, causing the sound to be 'clipped'.

Sound propagation is directional, of course, which further complicates matters. MEMS devices are omnidirectional, picking up sound from all directions equally. This is desirable for some applications, but not others, such as when we wish to isolate a specific signal and minimize sounds coming from elsewhere.

One way to make a microphone directional is to collect more of the signal from a specific direction and feed this excess into the transducer. Sounds from other directions are still present at their normal levels, but they are now *relatively* weaker. This is the idea behind a reflecting, or parabolic, dish.[16] The large cross-sectional area intercepts a larger portion of the ever-spreading longitudinal waves than the transducer alone could. The parabolic shape—which we saw in Chapter 1 is a conic section—ensures that waves coming 'straight in' along its axis reach the transducer, while sounds from other directions enjoy no such collimating effect, but are scattered this way and that.

Making the dish larger improves the directional amplification, but also mitigates an inherent limitation of the dish: namely, its poor low-frequency performance. The parabolic dish's shape reflects waves from anywhere on its surface to come to the same focal point, provided they arrive moving parallel to the dish's axis. Astronomical telescopes exploit this principle, effectively acting like folded-up versions of a magnifying glass, taking in rays of light with wavelengths some million times smaller than the mirror and reflecting them all to a single place. But audio wavelengths can be large, comparable to or exceeding practical dish dimensions. For example, calls from a Horned Guan *Oreophasis derbianus* with frequencies under 140 Hz have wavelengths over two meters long.[17] When waves enter an aperture comparable to or smaller than their wavelength, they will spread out, or *diffract*, as they pass through.[18] This spreading negates the dish's collimating effect, and the reflections no longer come to the same focal point. A dish one meter across, already unwieldy, provides no gain below 340 Hz. Above that, the performance will steadily improve as the frequency goes up.

Since a large parabolic reflector can be impractical for field use, other more compact designs that achieve high directionality are more popular. An alternate strategy is not to boost the forward signal, but to attenuate the sounds from other, 'off-axis' directions. This is done by giving off-axis signals multiple paths to the transducer and exploiting destructive interference effects.

Typical directional microphones, such as the cardioid or supercardioid, enclose their transducers within asymmetric chambers appropriately called an 'acoustic labyrinth'. The front is open, while the back and sides feature multiple entry points and a maze-like interior between the apertures and the transducer. Waves from the side and back enter at various locations, travel different distances, and interfere destructively.

A similar approach that is a bit easier to visualize is the *interference tube*, which is the basis for shotgun microphones.[19] Figure 6.3 (a) shows a transducer sitting toward the back of a tube that is open at the front end and has parallel holes along its length. Sound arriving off-axis causes the air at each hole to act as an individual wave source. In the simplified example in (b), the spacing

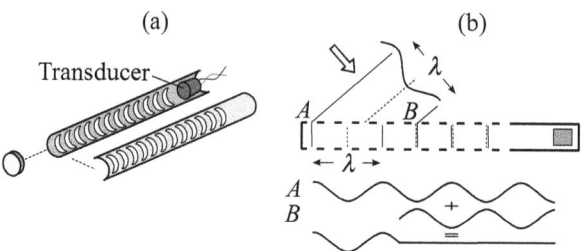

Figure 6.3 (a) Simplified schematic of a shotgun microphone interior, with the cover removed and the interior opened. The transducer sits at the rear behind a series of regularly spaced openings. (b) Side view showing the effect of interference from two selected points along the tube, marked *A* and *B*. An off-axis signal with wavelength λ stimulates a horizontal wave at point *A*. The horizontal distance to point *B* is one-and-a-half wavelengths, but the off-axis distance is a single wavelength, leading to destructive interference at *B* and no signal arriving at the transducer on the right.

between the holes corresponds to half a wavelength, leading to destructive interference at the transducer.

The obvious problem here is that interference depends not only on the spacing of the apertures, but also on the wavelength of the sound. Low frequencies with wavelengths much longer than the path differences will see little cancellation. A longer (more unwieldy) tube is needed to overcome this. Just as was the case with the parabolic dish, we can only deal with the long wavelengths of low-frequency sounds by building a bigger device. As a practical matter, directional sensitivity will always be frequency dependent, and the off-axis sound will therefore be 'colored' or distorted, a trade-off for its lower overall level.

If the microphone has done its job well, it will output a clean, undistorted signal that varies in time in concert with the audio. Now fully in the electronics realm, it can be stored for posterity, as well as manipulated and analyzed in limitless ways.

Analog and digital signals

A quality microphone is a sophisticated device, but its electrical signal is no less transitory than the audio it replicates. Additional technology is needed to record its output, store it and play it back later. For many decades this was accomplished with increasingly portable machines that could 'write' and 'read' signals to and from a magnetic tape. Everything about the process exploits electromagnetism: the electrical signal passes through a coil to create a changing magnetic field that the tape moves past, imprinting it. It can later be replayed by using similar transducers two more times: first by moving the tape past a coil, so that its changing magnetic patterns induce a current; and again when an amplified version of this current is used to drive a loudspeaker, which uses an electromagnet to generate audio waves.

The magnetic pattern written onto tape varies smoothly as the spool unwinds, in lock-step with the current and audio that created it. All of these are examples of *analog* signals: that is, continuous changes in some physical property. Using analog copies of sound—whether for recording or transmission—is simple and intuitive, but also prone to corruption. A phonograph record, which uses a thin groove cut with a shape corresponding to the audio, is easily ruined by a scratch because it physically, irreversibly alters the signal. An analog AM radio transmission is similarly marred by environmental effects such as electrical storms. Magnetic tapes lose fidelity due to other unavoidable problems.

We can minimize such headaches by using the digital domain, in which a signal is approximated by a series of numbers. The mathematical properties of such signals can be exploited to make them far more robust than their analog counterparts.[20] Analog and digital representations of a signal are shown in Figure 6.4. The analog signal (a) is a continuous function of time. The digital version (b) approximates it by taking regular samples at some set sampling rate. In (c) we have a crude digital representation, which is just a series of amplitudes.

The process of sampling is performed by an analog-to-digital (A/D) converter: an active circuit that makes repeated amplitude measurements. It must do so continuously in real time, sensing and transcribing the voltage into numbers during the short time interval between successive samples. This leads to a key question: how fast should we (or can we) make these measurements?

The immediate answer might be 'as fast as possible'. Digitization means saving only portions of the original waveform and discarding the rest. If we sample too slowly, the record will be too sparse, and a poor approximation. On the other hand, there are diminishing returns (and practical limitations) to making the time between samples ever smaller. An A/D converter can only operate so fast, and what's more, the more samples we take, the more numbers we will have to store and play back later. The practical solution is to

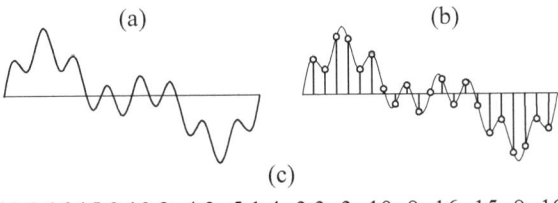

Figure 6.4 (a) An analog signal as amplitude versus time. (b) Digitization of the signal is performed by taking samples at regularly spaced time intervals. (c) The digital signal of measured amplitudes ordered in time is simply a list of numbers.

digitize just fast enough so that we don't lose any part of the signal that we could hear, but no faster.

If we don't sample often enough, we will pay a price by losing the higher frequencies, because their rapid oscillations occur between samples and escape detection. In order to 'see' a high frequency, we must capture at least two samples within its period; think of marking off both a crest and a trough. In other words, we need to sample at twice the highest frequency that we wish to preserve; this is *Nyquist's theorem*, and it is a bedrock principle of digital recording. The standard sampling rate historically settled on for audio is 44.1 kHz, which ensures that we lose nothing within our range of hearing that tops out around 20 kHz.

Every second of audio now becomes a list of 44,100 numbers, and now we must grapple with how to express and store the great many digits that the A/D converter will be spitting out. Digital electronics operates entirely in binary, as electrical states that are either on or off, 1 or 0. We can connect our familiar notion of numbers to such binary digits (bits) only by using strings of 0s and 1s, with the string length determining the range we can express. Three bits, for instance, can only be written down in eight unique ways: 010, 110 and so. It isn't hard to see that the progression goes as powers of two: 2, 4, 8, 16, 32, 64 and so on. With a series of N bits, we can represent 2^N different numbers.

N is called the *bit depth* and is a kind of amplitude analog of sampling rate. If it is too small, the digitized signal will have a certain 'blockiness' to it, and more importantly, a limited *dynamic range*: that is, the ability to capture both quiet and loud sounds. A large analog amplitude will exceed what our short string of bits can express. This results in the signal becoming saturated, or 'clipped', and often a noticeable loss of fidelity.

Typically, digital recordings have used a bit depth of $N = 16$ (in a CD, for example) or 24 (Blu-ray disc audio). A 24-bit audio signal provides a dynamic range that is larger than what the human ear can hear. Electronics with 32-bit resolution are becoming more widely available, and this is a game-changer for field recording, because the recordist need never adjust the amplification between microphone and recorder. There is so much 'headroom' that the loudest imaginable noise cannot be clipped; and there is so much resolution that faint sounds can be amplified in post-processing without becoming 'grainy', as they would for low-bit-depth signals. The trade-off, besides the cost for the more sophisticated and costly A/D converter, is the larger file size needed for all these additional bits.

A digitized signal, now just a long chain of 0s and 1s, is easily stored or transmitted. Perhaps more importantly, its mathematical character can be put to ingenious use: whenever we save or send digital information, some amount of *error correction* can be employed. Error correction, simply put, means adding additional bits to make the signal resistant to corruption. The methods for doing this can be quite complicated, but a simple example demonstrates the idea. We imagine a system where we send information back and forth in sets of seven bits. For each of these sets, we'll now include an eighth bit—a *parity bit*—with the following rule: if the seven bits represent an even number, the parity bit is zero; if they represent an odd number, the parity bit is one. Now, when we receive a block of eight bits, we first check it against our rule—if the seven bits express an even number, but the parity bit is one, we know that something was corrupted during transmission; some error occurred. We can then ask that these bits be sent again. This simple example isn't foolproof—two errors will go undetected, for example—but it helps reduce to the overall error rate and, despite its simplicity, it is still used in various applications. More advanced methods, including those used in QR codes, can pinpoint which bits were corrupted and fix them.[21] Some of the ubiquitous square codes that we scan with our smartphones may contain roughly twice as many error correction bits as actual data bits, ensuring that even with a quarter or more of the block obscured, it yet remains readable; try it!

Robustness is just one benefit of digitizing a signal. More important is that its numerical form makes it amenable to analyses with a vast set of digital signal processing tools. Of particular interest from an audio perspective is the separation of a complex waveform into simpler components.

In Chapter 5, we highlighted superposition: the simple way in which waves add together, giving rise to effects such as constructive and destructive interference. We also saw how adding multiple sinusoids could create repetitive patterns with complicated shapes. Any periodic waveform shape, in fact, can be built by adding up different sine waves: one of the loveliest results in all of mathematics. Far more impressive is that we can do the reverse process; starting with any wave, we can deconstruct it, finding all of the constituent sinusoids. The tools for doing this are part of *Fourier analysis*, which was formally developed by physicist Joseph Fourier and others in the late 1700s.[22] It is what allows us to make spectrograms, the informative visual depictions of audio signals.

Spectrograms are useful because they intuitively display the frequency content of a song or call, as well as how this content changes over time. All this information is in the original waveform but it isn't recognizable, because the frequency is revealed only by the rapidity of the oscillations from positive to negative, which is far too dense to be resolved unless we zoom in, in which case we lose sight of the entire signal. Because the original captured signal is a display of amplitude versus time, it is said to be in the *time domain*. To make a spectrogram we'll need to map out amplitude versus frequency: that is, to view the signal in the *frequency domain*. This all becomes more intuitive as we jump into the process, which we'll do next.

The frequency domain and spectrograms

Suppose we've recorded and digitized the short burst of audio shown in Figure 6.5, where we've zoomed in to show just the first few cycles. The waveform is periodic, but is more than a simple sine wave. It appears to have several components: sinusoids with different frequencies added together. We might estimate the periods and amplitudes by eye, but we want a rigorous way of quantifying it, something we can ask a computer to do. We'll now sketch out how Fourier analysis does this for us.

The strategy involves comparing the signal with a sine wave of a particular frequency and extracting a number, a coefficient, which tells us if that specific sine wave is part of our signal. Then we systematically repeat this process for a range of frequencies. We'll first show how we get a coefficient, and then consider some other details.

In Figure 6.6, we have the audio waveform from Figure 6.5, and a sine wave that we'll compare it to. The key idea is to take the *product* of these two curves: that is, we multiply the numerical values of their amplitudes. The result is shown on the right. When both amplitudes are positive, or both negative, the product is positive; wherever they are opposite in sign, the product is negative.

Note that the product curve is symmetric about the horizontal axis; it extends above like a mirror image of the part below and if we averaged it up, it would come out to zero. The square of that average value is the coefficient

Figure 6.5 Several cycles from a periodic but non-trivial analog waveform of amplitude vs. time.

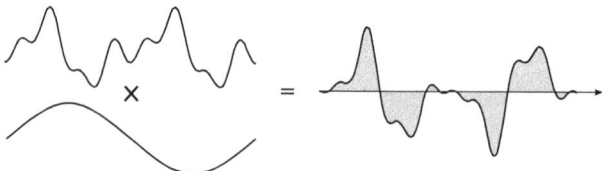

Figure 6.6 The original analog signal (top left) is multiplied by a sine wave with a wavelength twice that of the signal. The product on the right has equal area above and below the axis, and hence an average value of zero, indicating that this sine wave is not a component of the original signal.

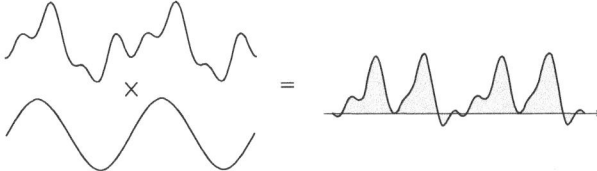

Figure 6.7 The original analog signal (top left) is multiplied by a sine wave with a wavelength the same as that of the signal. The product on the right is not symmetric about the axis, and hence its average value is non-zero, indicating that this sine wave is a component of the original signal.

we are looking for. It is zero for this case, because our waveform does not contain this particular frequency.

Now let's choose a different frequency, one that is present in our signal, as illustrated in Figure 6.7. When we multiply these curves, something new immediately jumps out; because they are mostly positive together or negative together, multiplication yields mostly positive numbers. When we then take the average of the curve on the right, we will be left with a number that is not zero. The size of this number will reflect how much of our original signal consists of this frequency.

Note that the phase of the sine wave in Figure 6.7 happens to line up with the original signal. If we had shifted it, using a cosine (a sine shifted by half a wavelength) instead of a sine, this procedure would have yielded a zero. In practice, we'll use both sines and cosines and add the resulting coefficients in order to account for any differences in phase.

Our signal contains another, higher-frequency component that did not contribute to this coefficient, because its contributions were equally positive and negative, as is clear from Figure 6.7. But when we repeat the procedure using its frequency, we will get a different non-zero number, providing the coefficient for *that* frequency.

In broad strokes, the procedure is to repeatedly choose a frequency, create a sine wave, multiply it against our signal, and obtain the coefficient; but which specific frequencies should we use? Since we've already digitized at a sampling frequency f_s, there is a natural set of frequencies that recommend themselves: namely, $f_s/2, f_s/3$, and so on all the way down to a lowest frequency of f_s/N, where N is the total number of samples in the signal.

Once we've calculated all the coefficients, we can plot them versus frequency to display a spectrum, or frequency-space representation. Figure 6.8 shows examples of waveforms and their spectra. Figure 6.8 (a) is for the example that we have been working through, which had three components. In (b) we show a square wave, which breaks out into many components. Signals that have 'sharp' edges or steep slopes can only be built by using high frequencies.

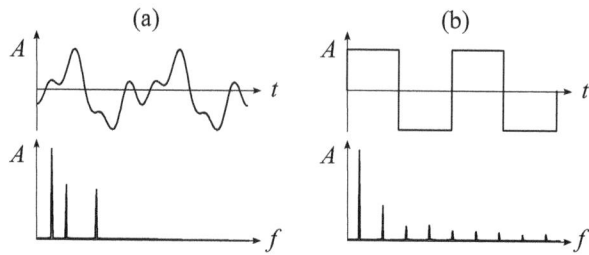

Figure 6.8 (a) An analog waveform in the time-domain (above) and the corresponding spectrum (below) showing amplitude versus frequency. (b) A square wave and its spectrum, which contains many more high-frequency components.

This process is the discrete Fourier transform (DFT). In the example worked through above, we used only a short length of a signal for illustration purposes. What happens if we apply this technique to more than a tiny snippet of sound? An audio signal such as a bird song is far more complex than a few sinusoids added together. It begins with silence, and may be punctuated by syllables and changes in volume, pitch and timbre, before finally returning to silence. The variation, even within a brief burst, can be considerable, but for the sake of illustration we'll consider something not too daunting: the two-part 'Hey Sweetie' song of a Black-capped Chickadee *Poecile atricapillus*, shown in Figure 6.9 (a).

Running the DFT on the entire song produces a spectrum with two clear peaks, shown in (b), that captures both the first high note and the lower second note. But this spectrum cannot tell us which note came first, or anything else about how it changes over time. If we were to break up the audio between 'Hey' and 'Sweetie' and create spectra for each part, the first half would contain just the high note, and the second only the low one, as shown in Figure 6.10. This suggests how we might want to proceed: a long recording might be sliced up into a series of shorter recordings, with the DFT analysis run on each one.

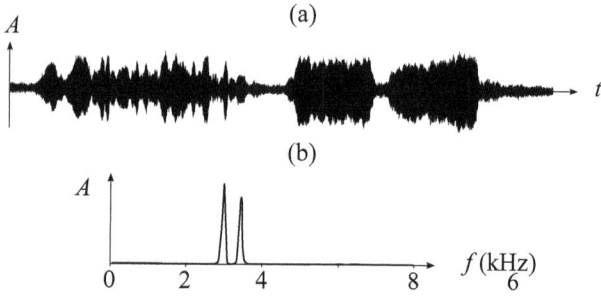

Figure 6.9 (a) Time-domain signal for a full Black-capped Chickadee song. (b) The corresponding spectrum for the entire signal, showing two main peaks between 3 and 4 kHz.

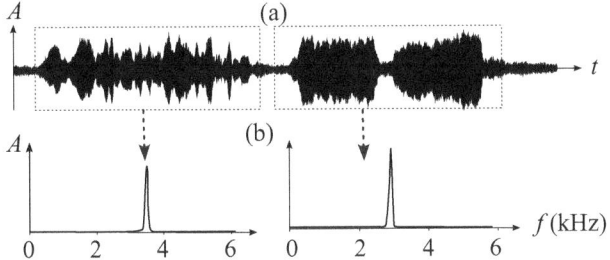

Figure 6.10 (a) Time-domain signal for a full Black-capped Chickadee song, split into two sections. (b) The spectra for each section—each now with a single, different peak frequency—corresponding to the 'Hey' and 'Sweetie' notes.

This will increase our resolution of how the spectrum changes over time, but also presents us with more spectra to look at all at once.

Such a series of plots quickly becomes overwhelming. The brilliant solution is to use color or gray scale to represent coefficient amplitude; to place frequency along the *y*-axis, so that going upward naturally represents going upward in pitch; and to have time flowing left to right along the *x*-axis. This results in the familiar spectrogram, as shown in Figure 6.11. It is simply a series of spectra running vertically and placed side-by-side.

How small should we, or can we, make these partitions? Some care is needed whenever we apply an analytical knife like this. We want fine enough divisions to avoid a 'choppy' image, but we need to keep in mind that more partitions increase the amount of work and file size. Moreover, as we zoom down to smaller time scales, we lose our ability to see long wavelengths, or low frequencies. If each chunk of signal that we analyze has a length t, the DFT will only be able to capture frequencies as low as $1/t$. To get a feel for

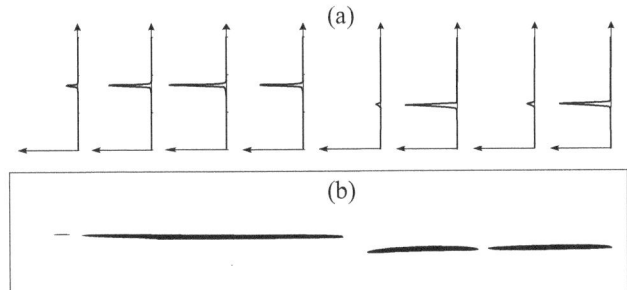

Figure 6.11 (a) Individual spectra taken after partitioning the Chickadee song into eight sequential sections. Here, the plots have been rotated on their sides, with the frequency axis pointing up. (b) The spectrogram for the entire song, which displays the amplitude in the space of frequency vs. time. Here, amplitude corresponds to the thickness of the line, but typical spectrograms will use a color gradient in the display.

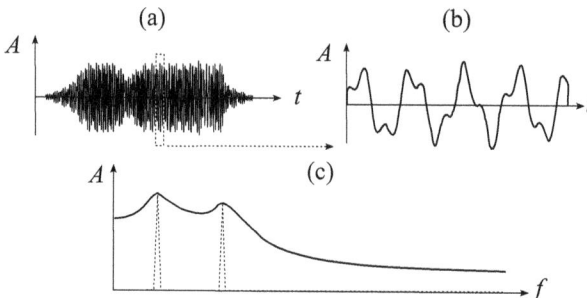

Figure 6.12 (a) A long signal to be analyzed with a STFT. (b) An example of a section partitioned from the full sample. Although it only consists of two sinusoids, the partitioning has left the ends discontinuous. (c) The resulting spectrum is artificially inflated relative to the two single peaks that we would expect to see, showing the 'spectral leakage' caused by the process.

this, suppose we divide a signal into blocks of 2,048 samples each. With a sampling rate of 44.1 kHz, these amount to frames 0.046 seconds long; and the corresponding frequency is about 21 Hz, near the lower threshold of hearing. This would be a sensible choice for audio.

This strategy of first dividing our audio into partitions and then running a DFT on each one, is called the short-time Fourier transform (STFT) and is the method behind every spectrogram. This still isn't the end of the story, though, because subdividing the signal introduces unwanted artifacts. Consider the waveform in Figure 6.12 (a). In order to apply a DFT, we need to first partition this into a series of shorter signals, such as that shown in (b). A problem is that the DFT will treat the 'sharp edges' at the start and end of this segment as being part of the signal, but they are only artifacts because of how we chose to cut it. The resulting spectrum (c) will not be accurate, as it is suffering from 'spectral leakage', a kind of smearing out. Even if only two frequencies (shown in dashed lines) truly existed in the original signal, the spectrum won't be sharp, but will indicate frequency components which are not present.

We can mitigate this by using a *window* when we partition the signal. A window is a smooth function with the same width as the partition, but with the important property that it gently tapers to zero at the ends. Figure 6.13 (a) shows the commonly used 'Hann' or 'Hanning' window. When this function is multiplied against the original sample, shown by the dotted line in (b), most of the original shape is preserved (solid line), but the sudden changes at the ends are gone. The spectra with (solid line) and without (dotted line) are shown in (c); the application of the window accentuates the peaks and reduces the smearing from spectral leakage.

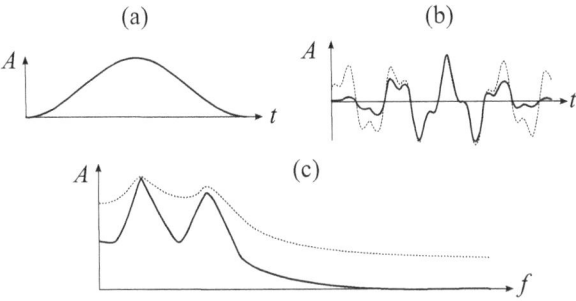

Figure 6.13 (a) The Hann window. (b) Application of the Hann window is performed by multiplying it against a section of the partitioned signal. The dotted line is the original, the solid line is the product, which now approaches zero smoothly at the ends. (c) Spectra with (solid) and without (dotted) the Hann window applied, showing how the spectral leakage is reduced.

This all requires an enormous amount of calculation. Running an STFT on our audio is feasible only because we can take a shortcut via what is arguably the greatest computer algorithm ever developed, the *fast Fourier transform*, or FFT.[23] The FFT yields the same results as a normal DFT, but far more rapidly, by exploiting symmetries of sine waves in order to reduce the number of steps. For a block with 1,024 samples, for example, an FFT is a hundred times faster than a full DFT. Without this, we might be waiting for the better part of a minute for a spectrogram to appear on our smartphones.

In the end, a spectrogram is merely a two-dimensional array of numbers: like a vast spreadsheet of rows and columns, where each cell holds a value, or gray-scale number. This means that any two images can be quantitatively compared to one another to see how alike they are, by measuring how well the values in these cells correlate. We might do something similar in spirit to a Fourier analysis—multiplying the two images together, cell by cell, and adding the results—to provide a measure of similarity. In this way, birdsong audio can be compared in a systematic way to a series of references in order to find the best match. This is the basic idea that enables applications such as Merlin to make accurate identifications.[24]

Audio recording in the field has been going on for a century, but the digital tools are merely a few decades old. They have become so inexpensive and portable so rapidly that we overlook how astonishing and powerful the technology is. Our field observations can now be made far more impactful; with a modest investment and scarcely more effort, high-quality audio can be captured and submitted to an online library to become part of a permanent, valuable collection.[25] For a growing number of us, the dedicated harvest and preservation of bird song is as rewarding and addictive a pursuit in its own right as birding.

Chapter 7

At the Lake: Sunlight, Reflection and Refraction

Some bird outings begin with a pre-dawn arrival in the field, and we might use the time to ruminate on the coming sunlight, for its warmth, its illumination and its role as the primary enabler of life. In this chapter, we'll start with sunlight's origins and explore its properties as it streams down to reach an egret on the water. We'll introduce scattering, polarization and the fundamentals of reflection and refraction, which are critical to understanding color, vision and optics.

Sunlight and its spectrum

Because birding is typically a visual pursuit, much of the relevant physics concerns what we see and how we see it. The origins of avian colors, the ways in which our eyes and cameras form images, and the principles by which optical instruments manage to bring distant objects up close are the broad topics of the following three chapters. So here, we'll spend a few pages becoming familiar with some key properties of light. We've already covered some relevant topics, including the electromagnetic (EM) field and wave motion; and thanks to nature's economy, some concepts from the previous chapter on sound apply here as well.

The primary source of natural illumination is the same engine powering almost all biological and ecological processes. In this sense, nothing could be of more central importance to the naturalist than sunlight. Yet we easily take it for granted, and few of us ruminate on its origins, although we are so fortunate to have had the physics puzzled out for us. Even a century ago, the sun was a mystery, incorrectly understood; but today we possess knowledge that geniuses like Newton, Darwin and Faraday would have paid dearly to have. An account of solar physics is beyond what space will allow here, but a few remarks are in order.

When we learn that the sun is primarily hydrogen—the explosive fuel that reduced the *Hindenburg* to ashes—it accords with our inclination to think that it 'burns' like the propane-filled mantle of a camp lantern. But burning is a chemical reaction, and the solar interior is far too hot for molecules to even exist. Hydrogen generally consists of a single, massive proton bound with a much lighter electron but, heated sufficiently, it becomes *ionized*, which means that the electron is shaken loose. In the solar core, this ionized hydrogen is incredibly compactified—it is ten times denser than lead—and incredibly hot, around ten to fifteen million degrees, something that is simply beyond our capacity to grasp or visualize.[1]

The positively charged protons fiercely repel one another, but their extreme speeds associated with the high temperature permit them to get so close that the attractive, short-range strong nuclear force—which we can think of as a kind of subatomic Velcro—is almost sufficient to allow the particles to cling together. For this to happen, their electrical repulsion must be mitigated, and there is an obscure mechanism for doing just that.

A proton consists of two 'up' and one 'down' quark, both of which couple to another fundamental field: the *W+ boson field*. This is critical, because it permits an up quark (charge +$\frac{2}{3}$) to transform into a down quark (charge –$\frac{1}{3}$) and a W+ boson (charge +1), as the total charge remains the same. The W+ boson is unstable and soon decays into two other esoteric particles that won't concern us. What is important to our purposes is that, because of this coupling, a proton can become a neutron (one up and two down quarks). With the electrostatic repulsion removed, a newly formed neutron can eagerly pair up with the proton via the strong force to form a *deuteron*, releasing energy in the process.[2]

The crucial piece here, the interaction between the quarks and the W+ field, is an example of the *weak nuclear force*. We think of forces as pulling or pushing: gravity pulls us down, magnets attract or repel, and the strong force acts like an adhesive. The weak force has no such analogy; it is rather an *interaction* that couples particular fields together, facilitating the metamorphoses between certain particles. It's a subtle, critical feature of nature that we could not exist without; the energy production of the sun depends upon it.[3]

Within several seconds, typically, the deuteron will pick up a second proton, and we'll now have a new element: *helium*, in its He^3 isotope form. As the second proton snaps into place, even more energy is released. This new nucleon will then wander through the core for some millions of years on average, until it collides with an identical He^3, leaving the two neutrons and two of the protons clumped together—a He^4 nucleus—while the other two protons are sprung loose with energy to spare. In this fusion process, six hydrogen eventually produce one helium, leaving two hydrogen behind; this produces copious energy in the form of electromagnetic radiation, which keeps the temperature and pressure high enough to sustain the process.

The electromagnetic radiation released in the sun's core during fusion isn't visible light, but far more energetic *gamma rays*. If we were to follow a single gamma ray as it scattered off the teeming protons and electrons, we'd find it making a random walk (are we surprised?) from the core to the surface, encountering progressively lower pressures, densities and temperatures. At the visible solar exterior—the *photosphere*—the temperature is 'only' about 5,000 °C, comparable to that of a welder's electric arc, and the radiation largely consists of visible light.

It is well known that sunlight takes between eight and nine minutes to reach the earth, but the random walk that gamma rays take to move from the dense core to the surface requires between 100,000 and 1,000,000 years.[4] Given that numerous North American speciations are thought to have occurred in the last 250,000 years, some of the light we enjoy reflecting off some birds descends from radiation that began its journey out of the sun's core long before the species even existed.[5]

Although the fusing nucleons produce a staggering amount of energy, it is overwhelmingly 'wasted', spewed out into empty space, illuminating nothing. Because of the earth's distance and (relatively) small size, it collects only about 50 parts per billion of the total sunlight produced. And for as much as the sun dominates the daytime sky, it is yet of (relatively) diminutive form. Sharing the approximate angular size as the moon, it is only about 0.5° wide. Comparing it to the entire vault, horizon to horizon, it takes up only about one part in 100,000 of the sky's area. It isn't inaccurate to consider it a 'point' source of radiant energy.

In the previous chapter, we looked at the production of sound, where vibrating membranes coupled into waves, which propagated out through a medium, heading off in every direction. The medium of air allowed these waves to reach and couple to other objects, including various types of transducers that permitted the signals to be detected. The situation is similar here. The distant sun, relatively small in extent, comprises oscillating charges, which drive waves that radiate out evenly. The medium carrying this energy is the electromagnetic field, and when the waves couple with charged particles, they will vibrate in concert. Various kinds of transducers—such as eyes and sensors—exploit this coupling, allowing the light to be detected. Sound and light, so unlike one another, are built upon the same underlying simplicity. The inverse-square roll-off with intensity applies to both, and the idea of a spectrum is no less applicable here than it was for audio.[6]

Figure 7.1 shows the intensity, or radiance, of sunlight as a function of wavelength. In describing light, it is preferable to use wavelengths instead of

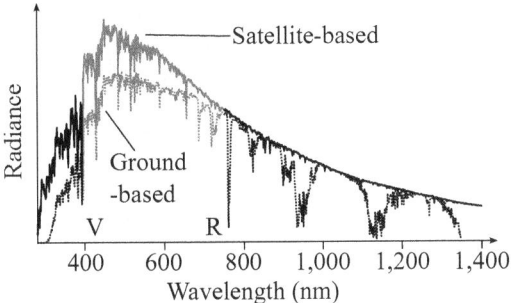

Figure 7.1 Comparison of solar spectra from earth and space, where the radiance is plotted vs. wavelength. The shaded region indicates visible light, with V and R marking the violet and red extremes, respectively. The heavy line is the spectrum measured from a satellite, free from the absorption effects of the atmosphere. The dashed line shows the spectrum measured from the ground. [Source data: National Renewable Energy Laboratory.⁷]

frequencies, as the latter are extremely high—over ten billion times higher than those involved in hearing—and difficult to reckon with. The shaded region indicates the visible portion, with the shortest wavelengths corresponding to violet (V) and the longest to red (R). The heavier curve shows data from a satellite above the earth, while the dotted curve is ground-based. There is an obvious filtering in the earth's atmosphere, evident as a series of bands or gaps where radiation does not pass. These result from gases in the atmosphere that readily absorb particular wavelengths of radiation.

The solar radiation with wavelengths less than 380 nm is primarily ultraviolet (UV) light, which we will discuss below. UV's reputation for having destructive effects stems from the fact that short-wavelength, high-frequency radiation carries more energy. X-rays and gamma rays have shorter wavelengths yet, and so are even more damaging; happily, they are not created at appreciable levels in the sun's photosphere.

On the long wavelength side, the spectrum's tail lies in the infrared region, which is where effective heat transmission occurs. Here, we see several pronounced dips in the ground-based spectrum due to the presence of water molecules, which are wont to vibrate, thereby absorbing energy and filtering out portions of the spectrum. These molecules possess a variety of ways to absorb energy. Douse them with EM radiation at microwave frequencies—which correspond to longer wavelengths, on the order of centimeters—and they will rotate in response. This is how a microwave oven operates, with the energetically spinning water molecules heating up the food that contains them.

Some filtering occurs in the sun's atmosphere, too, which is why irregularity appears in the spectrum taken from space. There is a fine structure—too small to see here—resulting from light being absorbed by atomic electrons, which then shift in their energy levels. Every atom and molecule has a unique set of electron energies, so spectra reveal which elements are present in the sun's outer layers. We'll explore this in Chapter 8, when we unravel how pigments give rise to various feather colors.

Skies and polarized light

Our atmosphere largely consists of nitrogen molecules: dumbbell-shaped, neutral atomic pairs whose six shared electrons make a strong bond. Although not inclined to interact with other gases enveloping the planet, N_2—like any molecule or atom—is not impervious to the effects of light. Exactly how a molecule responds depends on its electron configuration and the light's frequency. Atomic and molecular electrons are confined to particular orbitals, which correspond to specific amounts of energy. If light carrying the requisite energy impinges on a bound electron, it will be absorbed and the electron promoted to a higher orbital. The energy that light possesses is proportional to its frequency, so only light of specific frequencies can be fully absorbed in this way. Some sunlight will possess such specific frequencies, but most of it does not, and so avoids such absorption. However—and this is critical—the electrons will yet be affected; the light's oscillating electric field will jiggle them a bit, causing them to generate their own EM waves in response.

This interaction of sunlight and molecular nitrogen is an example of *scattering*: a broad category of phenomena in which incident light is re-directed by the presence of matter. Scattering often changes certain qualities in the affected light, and in the case of the atmospheric scattering, there are two important results to highlight. One of them, which we will cover in the following chapter, is the preferential scattering of shorter wavelengths, leading to the blue color of the sky. The other is *polarization*.

In trying to picture the wave nature of light, we might imagine something like a ripple on a pond. In that case, gravity enforces the wave's only possible orientation: with the crests vertically higher, and the troughs lower, than the nominal water height. But for light, which ripples through the electromagnetic field, there is nothing special about the vertical direction. The electric field must oscillate perpendicular to the wave propagation direction, but there are various ways in which this can be done, as Figure 7.2 illustrates. In diagram (a) we have a light wave propagating to the right along the z-axis, with the electric field E oscillating along the x-axis. In (b) we depict this same wave, as seen 'head-on' by an observer located on the z-axis; the circled dot indicates that the wave propagates toward the observer. In (c) the orientation of E has been rotated by an angle α. In both cases, we say that the wave is *linearly polarized*.

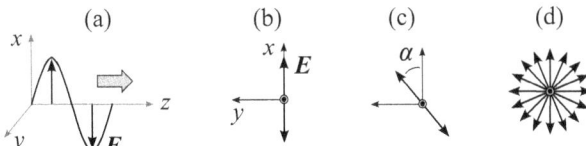

Figure 7.2 (a) The oscillating electric field **E** for a light wave moving along the z-axis, polarized in the x-direction. (b) Depiction of this wave, as seen looking down the z-axis. (c) Another linearly polarized light wave, inclined at an angle α. (d) A depiction of unpolarized light, composed of waves with a variety of polarization angles.

All that differs is the specific polarization angle. We now imagine the situation in which a collection of light rays with a variety of polarization angles are all moving along the z-axis. As one observes such light approaching, they'll find the electric field changing: not just along one direction, but in *every* direction in the x–y plane, as shown in (d). This is *unpolarized* light, equivalent to a uniform mixture of polarized light with many polarization angles.

With the sun's photosphere consisting of a jumble of charged particles oscillating in every which way, the incident sunlight reaching the earth is unpolarized. But when it strikes and scatters from a gas molecule, polarized light will be produced. Figure 7.3 shows a schematic of the process. Again, the light moves again along the z-direction, with the electric field oscillating in the x–y plane. Upon striking an N_2 molecule, the electrons must oscillate in the same direction that the field is changing, which means that their motion is also in the x–y plane, as shown with the arrows. These moving charges in turn create new waves, which radiate outward. When viewed from the side along the x-axis, as shown in the figure, we see that the **E** field for the scattered light can only be the y-direction: there is no way for a z-component to present. The presence of scattering molecules, and the geometry of the scenario, inevitably produces polarized light.

The polarization state of light isn't discernible (to us) unless we employ some device which makes it apparent. A 'polarizer' or polarizing filter reveals

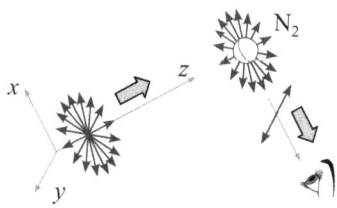

Figure 7.3 Unpolarized sunlight along the z-direction is incident upon a N_2 molecule, whose electrons must oscillate in the x–y plane, parallel to the electric field direction of the incoming radiation. Light that is scattered downward at right angles toward a viewer is polarized because it can only have a field oscillating in the y-direction.

such a state by exploiting material properties that allow the passage of light polarized along one direction but not another. To visualize this, imagine how the bars of a jail cell will only allow a long stick to pass through if it is oriented vertically. Hold up a pair of polarized glasses in front of a clear sky and rotate them, and the amount of light passing through will change dramatically. If the light were not polarized, there would be no such effect.

Beginning with studies of bees in the late 1940s, in which the direction of an insect's 'dance' was seen to change to match the orientation of the polarized light bathing it, biologists have shown that many animals—including various terrestrial and aquatic insects, crustaceans, cephalopods, reptiles, fish and birds—possess an innate ability to detect polarization.[8] For birds, the details are not entirely understood, but it has been observed in connection with magnetic field sensing and is consistent with the physics of cryptochrome, which we discussed in Chapter 4. Recall that the generation of a radical pair, which kickstarts the entire process, requires the absorption of light. Because of the protein's complex structure, some light polarizations produce the radical pair more readily than others; this means that polarization affects the detection of the magnetic field. Brilliant experiments on Zebra Finches' *Taeniopygia guttata* ability to orient and find food sources when the magnetic field direction and light polarization were varied independently confirmed that changes to either could cause disorientation.[9] Impressively, birds can respond to such trickery through a process of 'recalibration', in which conflicting cues are sorted out. A study on Savannah Sparrows *Passerculus sandwichensis* demonstrated that the birds will use sunrise and sunset in order to reorient themselves, not by referring to the sun's position, but rather the band of most strongly polarized light, which meets the horizon at 90° from the sun.[10] This is advantageous, as the sun's location is liable to be misjudged due to geological or topological variations. The work suggests that the birds may average their results from sunrise and sunset, making for an accurate and robust discernment of true north.

Polarization also occurs when light is reflected, generally speaking. This is commonly encountered when sunlight glints off the water and makes us thankful for polarized glasses. We turn next to the topic of reflection, and will say more about the polarization it causes later.

Reflection and least time

Our experience suggests that the physics of light and matter can become quite complicated, because we see such a variety of effects around us. Some materials—gases, and various liquids and solids—allow light to pass through them in a seemingly effortless manner. In other cases, matter can reflect in a

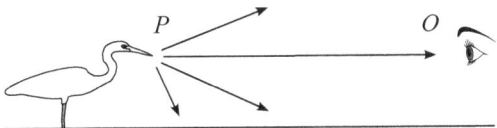

Figure 7.4 An egret in still water and an observer. Light from the point P radiates in various directions, including toward the observer at point O.

mirror-like fashion, effectively making a carbon copy of the light, but traveling in a different direction. A sheet of white paper also reflects, yet appears nothing like a mirror. A liquid such as milk may also allow some light to pass through, but not in a way that can maintain an image. Some materials will absorb certain portions of the spectrum but not others, resulting in a colored object, such as the leaf that takes in everything but green light. Our atmosphere will selectively scatter blue light, allowing other colors to pass through. Under some circumstances, matter can give us especially pronounced colors, such as when light is dispersed by a prism or the glow of iridescent feathers.

In each example, everything boils down to events at molecular and microscopic scales. Light waves push around charged particles, which in turn create more EM waves, and the visible results accrue via superposition. It is because the myriad charges making up matter can be arranged in so many ways that such a variety of effects are possible.

The case of light reflecting from a flat surface is obviously of fundamental concern, as the results are both straightforward and important. Before digging into the question of 'how, exactly?'—which involves mechanisms at the molecular level—let's get familiar with the rules that describe what happens.

Figure 7.4 illustrates a distant egret standing motionless in still water, seen by an observer. For simplicity, we will focus first on just a point—the tip of the bird's bill, marked as *P*—and consider what happens as sunlight bounces off to radiate outward. Some of that light will travel along a straight line directly to the observer at *O*. Some will strike the water and be reflected, also reaching *O*, but obviously not along a straight course. What path will the reflected light take, exactly?

This can be answered by going down to the fundamental level of interaction between EM waves and electrons, and we'll look at such an approach later. But first, it is helpful to employ an elegant principle that nature is fond of, referred to as the *principle of least time*; this idea is best explained with the aid of a diagram.[11]

Figure 7.5 (a) shows a simplified version of our scenario. We are specifically interested in how light from *P* can reach *O* by means of reflecting off the

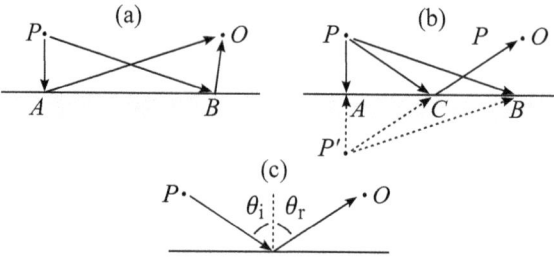

Figure 7.5 (a) Source P and observer O, with light taking proposed paths that involve points A or B. (b) Various paths taken from points P and its mirror image P', which must have the same length. Clearly, the shortest path from P' to O must go through C. (c) The correct path is such that the angles θ_i and θ_r for incident and reflected light, respectively, must be the same.

water's surface. We illustrate several different paths and consider if the light *might* take them. In one example, it moves straight down to the point A and then proceeds to O. A path with reflection at B represents another possibility.

For any path we pick—such as P to A to O—we can calculate the distance the light must travel and because it has constant speed, the time this path would involve. We might come to the diagram with a ruler and measure the distance, or determine a formula to capture these path lengths. A more inspired approach is to note that the paths made by P and a mirror image of P, called P', must have the same lengths, as shown in diagram (b). The shortest path from P' to O is a straight line, which passes through a point we denote as C. This means PCO must be the shortest path the reflection can use, and the angle at which light is reflected must be the same as the angle of its incidence. The custom is to measure the incident and reflected angles θ_i and θ_r relative to a line perpendicular to the surface, as shown in (c), and so write the rule as $\theta_i = \theta_r$.

We did this with a single point; but every part of the egret, or any other object, can be considered a point source that will have its own reflection. In this way, an entire image is formed. When rays of light that are parallel as they reach the surface remain parallel after they bounce off, we say that the reflection is *specular*. But if the water isn't flat—perhaps because of ripples passing by—the image will be distorted.

It is worth thinking about rippled water and its effect on the image. If the scale of the distortion becomes finer—if the wavelength of the ripples is made smaller—then the image will be broken into more fragments. Eventually, it will become so choppy that the object can no longer be discerned in the reflection, and only a vague blob of light remains. The rays of light that come in parallel do not stay parallel, and the reflection is not specular but *diffuse*. The ray diagrams for both cases are illustrated in Figure 7.6.

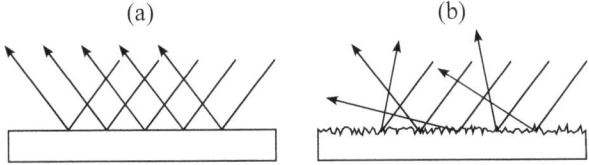

Figure 7.6 (a) Specular reflection occurs with a smooth surface in which all light is incident at the same angle. (b) Diffuse reflection occurs with a rough surface in which the incident angle changes randomly with position, leading to a wide range of scattering angles.

This is a simple example of how the microstructure of a surface can change its appearance, even though the composition of the material has stayed the same. Both a perfectly flat pond and a mass of whitewater crashing down a river are made of the same molecules and employ the same fundamental law of reflection, but only the former can act as a mirror. The whitewater has been turned into countless tiny reflecting surfaces oriented every which way, while the flat pond keeps them all perfectly parallel. In Chapter 8, we'll see how similar effects from microstructure will play a role in producing certain feather colors.

The mirror-like surface of tranquil water may appear like the metal-coated surface of a looking glass, a lovely coincidence given that its composition is so different. Unlike a metal, the water also permits some portion of the light to pass through, illuminating the fish below, which hold the attention of the egret. Next, we will look at how this happens, as well as the associated polarization of the reflected light.

Refraction and 'slow' light

The speed of light is the most famous constant of nature, the upper limit to how fast anything can possibly move. Denoted as c, its value is nearly 3×10^8 m/s in empty space. When it travels through those arrangements of matter that permit its passage—that is, through transparent materials such as gases, liquids or varieties of glass—it is affected in a curious way: it will require more time to propagate through the material than through the same span of empty space.

One might be tempted to make the following analogy. If we fire a bullet at a block of wood, its speed will drop as it penetrates the material. We might think light slowing in water or glass arises from a similar effect. Yet if the bullet should exit the other side of the block, it will move through the air more slowly than it did before impact. It could not re-emerge at the same high speed with which it initially struck, as that would violate the conservation of energy. Light, on the other hand, does exactly that: in striking a pane of glass, it arrives at c, progresses at a slower speed in the material, and jumps back to c after it streams out the other side.

This might seem a paradox, but it is explained by a different (seeming) paradox: namely, that the light within the material never actually slows down, even though it takes longer to traverse the material than it 'should'. Waves in the EM field always travel at c, but their energy gets coupled into the motion of the electrons they encounter. Each oscillating electron will in turn produce its own waves, which will radiate outward and affect all the other electrons, causing more oscillations in response, and so on. This looks like it will get hopelessly complicated, but when all these waves are added, they produce an elegantly simply result: that of a single wave moving at a reduced rate, even though the light moving between any two electrons in the material moves at c.[12] If that doesn't seem intuitive, we might simply leave it at this: the slow transit of light through a material is the 'cost' of having to set all those electrons in motion.

The same oscillation of charge in the material is also responsible for the reflection. Light doesn't 'bounce off' a surface in the way that a rubber ball rebounds off the floor. It is a re-radiation of energy because of the electrons' response to the light penetrating the material. We'll see shortly how this leads to polarization.

The effectively reduced velocity is given by c/n, where n is a dimensionless number—the *index of refraction*—which is always larger than 1. For air under typical atmospheric conditions, n is about 1.0003, giving a speed of about 99.97% of the vacuum value. For water, the index is approximately 1.33, causing light to effectively travel at 75% of its normal rate.

This slowed passage of light also causes *refraction*: the familiar 'bending' often observed at the interface of air and water. The importance of this effect cannot be overstated. Without it we could not make lenses that redirect light, nor construct optical instruments such as binoculars. We will see how refraction is exploited by devices ranging from eyes to telescopes to cameras in Chapters 9 and 10. For now, we'll concern ourselves with the rule that describes its operation. An elegant way to approach this uses the same principle of least time, introduced above. Figure 7.7 (a) shows a situation with a source of light P located in a vacuum (or in air, since the speed is barely changed), and an observer at O, located in a region with index n. Several candidate scenarios for the path the light might take are drawn for our consideration.

It is common to employ an analogy here: we'll take this image to be a bird's-eye view of a beach and ocean. A lifeguard on land at P sees a swimmer in distress at O, and must reach them as rapidly as possible. Since lifeguards

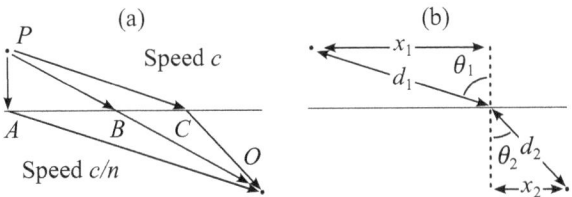

Figure 7.7 (a) Points P and O reside in regions with different speeds; c and c/n, and three different paths connecting them, are shown. These cross the boundary at points A, B or C. (b) A geometric construction for the optimal path, passing through C, which has the least time. The boundary is crossed at a horizontal distance x_1 from point P, which is a horizontal distance x_2 from O. Diagonally, the point of crossing is located d_1 from P and d_2 from O. The angle of incidence θ_1 has a sine given by x_1/d_1 and the refraction angle θ_2 has a sine of x_2/d_2.

can run across sand faster than they can swim, there is a question as to which path to take. The direct line is not optimal because it will involve a lot of swimming. Better to 'overshoot' the distance on land and spend less time in the water, which could mean entering at point C. The optimal strategy will depend on just how much slower the progress in the water is, which is captured by the factor n. If we calculate the fastest route, we will find it expressed by *Snell's law*, which states:

$$\sin(\theta_1) = n \sin(\theta_2)$$

Here, $\sin(\theta_1) = x_1/d_1$, and $\sin(\theta_2) = x_2/d_2$, as shown in Figure 7.7 (b). In the limit that the speeds are equal ($n = 1$), the angles are the same, which is to say that the light follows a straight line. At the other extreme—say the speed in the water is extremely slow—then the angle θ_2 will get closer to zero, minimizing the time spent in that medium.

Snell's law holds that—regardless of direction—the light moving from water (say, after being reflected by a fish) to the air will also undergo refraction. Depending on the point from which this light is perceived—perhaps by an egret—it can have an apparent position that differs substantially from its actual location, appearing higher than it really is, as shown in Figure 7.8.

Various studies have looked at how egrets and herons deal with this distortion in prey location. Birds that are mainly ground-feeders, such as Cattle Egrets *Bubulcus ibis*, will attempt to take fish when given the opportunity, but with less success than their piscivorous cousins.[13] Another species—the Reef Heron *Egretta gularis*—was found to select a particular orientation before striking such that a specific relationship between perceived and actual depth always held. When artificial barriers were introduced that prevented the herons from lining up in this preferred geometry, they were more likely to miss.[14]

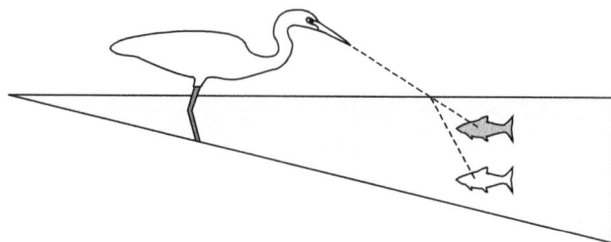

Figure 7.8 Light from a fish (white) is refracted upon entering the air. The apparent position then appears higher, as shown by the gray image, seen from the egret's perspective.

We might wonder if herons and egrets are challenged by another palpable effect we notice on the lake: the glare of reflected sunlight. Humans have only recently learned how to overcome this with the invention of polarized glasses, which function only because reflection—like atmospheric scattering—has a polarizing effect. Figure 7.9 illustrates why this is so, by focusing on a special case in which the reflected and refracted rays are at right angles. For simplicity, we assume the incoming radiation consists of two polarizations: the E field has components in the plane of incidence (indicated with arrows), and parallel to the surface (dots). Light enters the water and excites the response of molecular electrons, which move parallel to the electric field there, affecting the refracted ray and producing the reflection. In this case, the polarization direction in the water denoted with arrows is parallel to the direction of the reflected ray. Such an oscillation cannot produce a wave along that direction; that would entail E pointing along the propagation direction, which isn't possible. Hence, the reflection can only have in-plane polarized light. This effect is diminished at other incidence angles, but overall, reflected light polarization will have a pronounced in-plane character. Polarized glasses orient their material so as to block such light. Remove them, rotate them a quarter turn, and look through them at the water, and the glare will return.

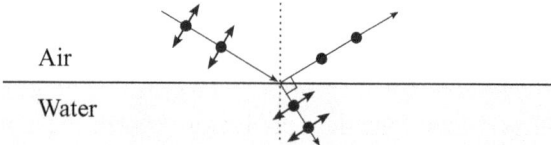

Figure 7.9 Light consisting of two polarizations, both in the plane of incidence (arrows) and parallel to the surface (dots), is incident upon water. In this case, reflected and refracted rays are at right angles to each other. The reflected light comes from oscillation of the water's electrons, which move parallel to the field. But below the surface, the component represented by the arrows is along the direction of the reflected ray. This cannot produce a transverse wave, so the reflection can only keep the in-plane polarization component.

A recent study concerned with the deleterious effects of reflections for herons searching for prey states that as these birds have no known anatomical adaptations that might exploit the polarized nature of the light, they might position themselves relative to the sun so as to minimize glint.[15] No such behavior was observed, although they cite evidence that it has been seen with pelicans and terns. Their suggestion is that some mechanism that exploits polarization may yet be discovered in piscivorous waders.

Glass, like water, permits most light to be transmitted; but the portion that is reflected is what endows windows with such a deadly allure for birds, making them appear as fictitious skies. While ornamentation can help prevent window strikes, it is encouraging that other solutions are being developed to maintain perfectly clear windows. This is because birds see UV light that we cannot, and novel treatments can add UV patterns that prevent collisions, while remaining invisible to humans.[16]

With the realization that birds could see UV, our notions about species that seemingly lacked sexual dimorphism had to be revisited. One study of 166 North American species with no male-to-female differences visible to human eyes revealed that over 90% of them had dimorphic features when imaged with UV-sensitive optics.[17] Much as some flowers reveal UV designs to insects and birds, but not to us, there are avian patterns—perhaps some of great beauty—which we unfortunately cannot appreciate with our native senses.[18]

This extended visual capability can also be of benefit to birds that hunt. Various predatory species, including Great Gray Shrike *Lanius excubitor*, Rough-legged Hawk *Buteo lagopus* and Common Kestrel *Falco tinnunculus*, exploit their UV vision to find prey.[19] The kestrel species in particular was shown to recognize the telltale signs left by voles, such as urine droplets, which are conspicuous to UV-sensitive detectors.

How and why do some materials reflect UV, or, for that matter, any other color? There are multiple ways in which different hues can be generated, and as the coloration of birds is so central to our enjoyment of them, we will spend the next chapter examining those details.

Chapter 8

During a Big Day: Light, Matter and Feather Colors

The species variety endemic to the peak of spring migration means an unrivaled variety of vibrant colors. For many, the summit of birding lies here, in the spectacular plumage of these living gems. In this chapter, we dive deep into the interactions of light and matter that generate this beauty. We'll see how structures comparable in size to the wavelengths of light can generate particular hues, while atomic and molecular physics underpin others.

Color and scattering

In the previous chapter, we traced how sunlight pours out from a superheated nuclear furnace, providing a spectrum of electromagnetic radiation, a small band of which our eyes have evolved to detect. Mixed together, the light in this band appears white, but thanks to a variety of interactions with matter, other colors are readily produced.[1] We'll now look into some of these mechanisms.

Physics has much to say about color, but we must also stress that color is a *sensation*: an inscrutable conscious experience. A laser with a wavelength of 532 nanometers will produce green light, but not every patch of verdant hue we sense has such an origin: a mixture of blue and yellow light is also *perceived* as green. And as with any color, 'greenness' as *qualia*—that is, as experience—cannot be expressed or explained. Moreover, demonstrations such as 'Benham's Top'—in which a moving black-and-white pattern causes the perception of nonexistent colors—show that our experience of hue is constructed within our heads.[2]

The physiological aspects of color perception will not concern us, beyond the minimum we need to know: we experience color because light has a particular wavelength (or frequency), or via the familiar mixing of colors that we learned about in our first art classes. Remove the red from a white light source, and what's left is bluish; mix the red with yellow and orange results, and so on.

Let us begin with the light in the sky. When the electrons in atmospheric nitrogen molecules are exposed to sunlight, they vibrate in response, re-radiating light. We've seen how this gives rise to polarization, but we didn't consider the role of the light's frequency in the scattering process. Recall that a bound electron's motion is restricted because it is held by the attractive force of nearby protons. It is not rigidly fixed in place, but acts somewhat like a weight suspended from a spring. If we gently tap the weight, it will bounce up and down; if we nudge an electron, the atomic forces act to restore it, but as with the spring (or a pendulum), the motion will 'overshoot' and an oscillation results at the system's *natural* or *resonant* frequency.

For a weight on a spring, this frequency will depend on its mass (larger is slower) and the spring strength (stiffer is faster). The minuscule electron mass, plus the considerable force from the nearby nuclei, together produce very high frequencies; for N_2, it is in excess of 10^{23} Hz. This frequency lies well above that of any light we can see; it corresponds to ultraviolet (UV) radiation. What we need to understand is how an electron will respond when exposed to visible light, which has lower frequencies.

Our situation is that of a *driven harmonic oscillator*, and how such a system responds depends crucially on the both the *driving* and natural frequencies. It is helpful to consider an analogy: standing in for the bound electron is a rocking chair, which naturally oscillates at some frequency determined largely by the curvature of its 'feet'. We could drive it into motion by placing it on the deck of a ship buffeted by slow ocean waves, which play the role of light. For simplicity, suppose the chair rocks at a natural frequency of 1 Hz. If the waves arrive at a lower frequency, say 1/10 Hz—it takes ten seconds to go from crest to crest—the chair will creak along gently in unison with the tilt of the deck. Should the waves speed up and arrive at 1 Hz, though, the chair will rock with greater and greater amplitude, eventually tipping over. This is similar to pushing a child on a swing; drive the swing at its natural frequency, and large amplitude will soon result, even with very little effort. As a general rule, the closer the driving frequency is to the natural frequency, the more energy can be coupled in and the larger the amplitude.

This means that the higher-frequency components in sunlight— corresponding to the blue and violet colors—will cause the electrons to oscillate with greater amplitude than will the lower frequencies from the red side of the spectrum. These electrons then act as tiny antennas, re-radiating

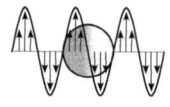
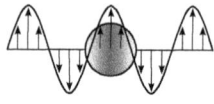

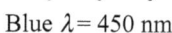

Blue λ = 450 nm   Red λ = 750 nm

Figure 8.1 A small water droplet 450 nanometers in diameter in the presence of blue (left) and red (right) light. The arrows represent the electric field polarization direction. The molecules within the droplet become polarized by this field, leading to a nonuniform distribution charge, indicated by the shading. As these molecules re-radiate, there will be no net radiation of blue light at a distance because half the sources are out of phase. For the red light, the polarization is more uniform over the droplet, allowing for some re-radiation.

the incident sunlight, but with a strong preference for the higher frequencies. This effect, known as *Rayleigh scattering*, is what colors the sky blue.

Individual H_2O molecules in the atmosphere respond similarly, but unlike N_2, they are attracted to one another and inclined to clump together. This is due to their bent shape, which distributes their charges in a non-uniform way. With sufficient numbers and at appropriate temperatures, they will condense into minuscule droplets of liquid, full of charges that re-radiate blue light preferentially. Should a droplet reach a size of several hundred nanometers or more, however, there are new effects that will alter this scattering.

To see why, first note that if a scatterer measures half a wavelength or so of the incident light, the charges on one side will feel the opposite electric field direction than those on the other. From a distance, very little or no scattered light emerges because half of it is out-of-phase with the other half, and destructive interference results. Expose it to a longer wavelength of light, though, and this effect disappears.

Figure 8.1 shows a comparison of a water droplet 450 nanometers in diameter relative to blue and red wavelengths of light. The blue light will drive the charges out-of-phase in a way that the red light cannot. Individual electrons still 'favor' blue light for its frequency, but the collective behavior favors red light for its wavelength.

A second, more important consideration is that as the droplet grows in size, the electric field that each charge experiences isn't just that from the incident light; there are so many other charges in the droplet, each of them producing their own re-radiated light, that the fields that *they* produce affect all the other charges as well. This superposition effectively blends the total response, such that the scattered light will have little color dependence.

What we are describing is *Mie scattering*, which occurs with spherical shapes on the order of a wavelength or more. Bathed in sunlight, a diffuse collection of Mie scatterers will re-radiate all the component wavelengths equally. Hence, vast collections of minuscule water droplets—clouds—produce white. The reds they present at sunrise or sunset come from low-angle sunlight that has been filtered by the Rayleigh scattering that sprayed the blue into our neighbors' skies, leaving only warm tones.

The cloud droplets that scatter light typically range in size from several hundred nanometers up to a tenth of a millimeter. As they grow larger they cannot be held suspended, and descend as rain. The electrons within these droplets have not lost their proclivity to oscillate more strongly in response to higher-frequency light, and this will produce another important effect: the droplet will act to slow the blue light more than the red. We've previously discussed how light in a transparent medium must acquire an effectively lower speed, which leads to refraction. Now we see that the degree of refraction is frequency dependent. Violet light will be 'bent' more than blue, which is bent more than green, and so on, upon entering and leaving a raindrop, ultimately leading to the *dispersion* of sunlight into a rainbow.

There are a dizzying variety of effects that light and matter can lead to, but the physical origins are simple at the core. Atoms and molecules will absorb selected frequencies, but mostly, their electrons act as driven oscillators that preferentially respond to higher-frequency (more blue) light. The light they re-radiate obeys the same superposition that all waves must, so it is the specific order (or disorder) of many molecules together that will determine what we see. With this in mind, we can explore the origins of feather colors.

Structural whites and blues

Clouds are opaque and white due to the cumulative effects of many independent scattering events, which require a considerable volume of material. A single cubic centimeter of cloud alone would be transparent, holding perhaps only 1,000 microscopic droplets ranging from hundreds to tenths of a millimeter in diameter. For something the size of a feather to appear white, the scattering surfaces will need to be far more dense, and in bringing them together, additional effects will arise.

If two reflecting surfaces are separated by an appropriately small distance, their combined re-radiated light will undergo interference: constructive, destructive or somewhere in between. From the perspective of a distant observer, interference is interference, whether it comes from two sides of the same droplet (as in Mie scattering) or the surfaces of two adjacent scatterers. If these separations vary appropriately in scale, the results will be washed out, contributing to an overall colorless (white) effect.

A foam is a perfect example of a relatively dense, disordered collection of variously sized scatterers. The volume of bubbles atop a freshly poured beer,

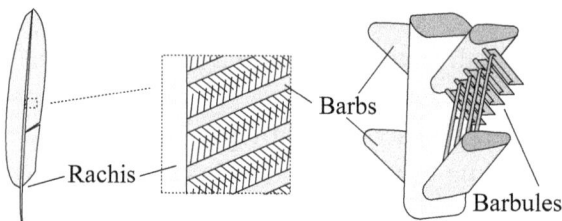

Figure 8.2 A simplified diagram of a flight feather on the left. In the center, we zoom in on a portion near the rachis in order to reveal the barbs which extend out from it. From the barbs extend on either side a series of smaller barbules. The close-up on the right shows how the barbules come in two varieties, one of which is hook-like and the other flange-like. This provides a Velcro-like structure to preserve the feather shape and strength. [Modified from Foth (2020).³]

or the rigid structure of cured polyurethane used for insulation or packing, appear white for the same reason. We remarked on diffuse reflection in the previous chapter, where disordered reflections were scrambled every which way. What we are noting here is that the wave superposition effects can be similarly jumbled. Some kind of disordered microstructure is responsible for every object that appears white when illuminated by sunlight, including white feathers.

Figure 8.2 shows a simplified diagram of a feather, in which a series of parallel barbs extend outward from either side of the central rachis or shaft, forming a flat vane. Each barb itself branches into barbules, which provide novel Velcro-like attachments. In most, but not all, cases, the color we perceive comes from the barbs.⁴

Feathers are primarily composed of beta-keratin: a structurally rigid, optically clear protein that also makes up a bird's bills, leg scales and claws. Feathers are strong and pliable, and yet also lightweight, because of air-filled structures mixed in with the keratin. While the details will vary for different types of feathers and species, the cross-section of a feather barb typically reveals a sponge-like interior made from air pockets that range in size, as shown in Figure 8.3.

The haphazard character of this air and keratin matrix that makes up the bulk of the barb leads to numerous, but ultimately disordered, reflections from the countless interfaces. This lack of regular arrangement in the ensemble produces *incoherent* scattering, which is wavelength-independent and hence colorless as a whole, because every color is scattered equally.

The absence of any structural uniformity or repetitive pattern in a foam might seem obvious from inspection. There is no apparent order, and the air pockets obviously vary in size. When blue feathers were first imaged at this scale, workers noted a similar structure of cavities that seemed to lack a strict order. Yet there was a subtle regularity present that was not readily discernible

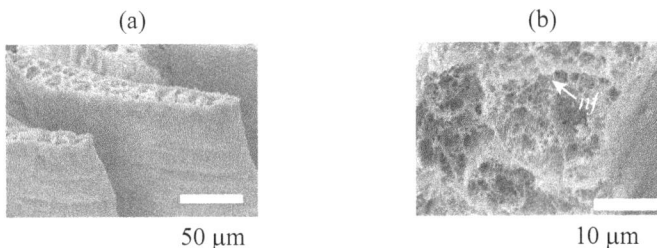

Figure 8.3 (a) Cross-section of a white barb, from the tail feather of a Eurasian Woodcock *Scolopax rusticola* imaged with scanning electron microscopy (SEM). (b) Further magnification of the barb reveals the disordered structure of the spongy interior and air pockets ranging in size. The term 'nf' in the image indicates 'networks of keratin fibers'. [Images from Dunning, Anvay, D'Alba *et al.* (2023).[5] Reproduced under the terms of the Creative Commons Attribution License.[6]]

from a cursory inspection of the images. The understanding of blue feather coloration was impeded for decades, in part for this reason.[7] A robust method to quantify the orderliness of the structure was needed.

It seems like a daunting task to measure and catalog the various sizes of the voids in a sponge-like structure, as well as the distances between them. Fortunately, there was already a tool which could easily do this, the same one we now use to analyze bird sounds and create spectrograms: the Fourier transform.

Recall that an audio signal involves the regular variation in sound intensity over time. When we apply the Fourier transform, the repetitive portions of the signal are captured and represented as peaks at particular frequencies. If we can apply this technique to a signal that changes in time, we can certainly apply it to one that changes in space. We could start, not with air pressure versus time, but with density versus position. In that case, the Fourier output isn't in terms of frequency (the inverse of time), but wavenumber, which is the inverse of length.

A black-and-white image of anything—including the keratin and air matrix in a feather barb—is simply an array of numbers, each indicating a grayscale value for a specific location, like points on a map. Each value is proportional to the amount of one material present at that place. One way we might approach the analysis is to break the image into a series of lines and look at them individually. This is illustrated in Figure 8.4, where both regular (a) and random (c) two-dimensional images are considered. In (b) and (d), we have selected horizontal lines from (a) and (c), respectively, one pixel in height, of the respective images, and plotted the grayscale values versus position.

In Figure 8.5, we take the one-dimensional samples and create their corresponding Fourier transformations, using the same fast Fourier transform (FFT) used for processing audio. Every point along the x-axes in the plots

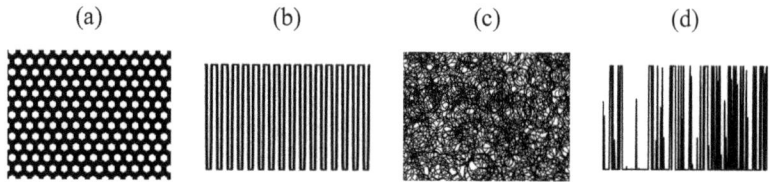

Figure 8.4 (a) An ordered two-dimensional structure. (b) A horizontal sample from (a), one pixel in height, plotted as grayscale value versus position, showing the same regularity. (c) A random two-dimensional structure. (d) A horizontal sample from (c), one pixel in height, plotted as grayscale value versus position, showing no regularity.

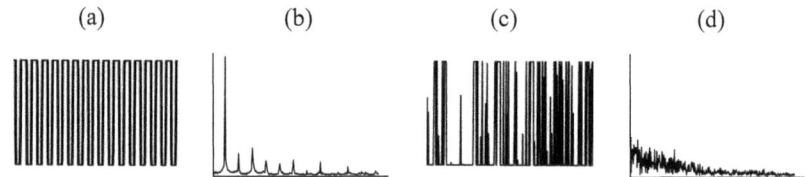

Figure 8.5 (a) Horizontal sample for the structured material of Figure 8.4. (b) Corresponding Fourier transform showing a clear peak. (c) Horizontal sample for the randomly structured material. (d) Corresponding Fourier transform showing no peak.

(b) and (d) represents a particular value of 1/length, and so also corresponds to a particular length scale. Interpreting these results tells us that the distance between interfaces for the ordered material has one primary value, indicated by the highest peak in the spectrum. For the disordered material, there is a no clear peak, but rather a range of separations.

For simplicity, we chose the central horizontal row of pixels; but we could have sampled points along any line within the original image for this exercise in order to determine how the distances are distributed along that direction. This suggests that we might scan the entire two-dimensional image and create a corresponding two-dimensional Fourier transform for a more complete picture of how the various length scales are distributed.

Figure 8.6 shows the results from this approach, with (a) and (c) corresponding to the same disordered and ordered images from Figure 8.4, respectively. The Fourier transforms in (b) and (d) can be read from the central point outward. For the ordered material, the two-dimensional spectrum in (b) has an obvious geometrical character, with points instead of the peaks of a one-dimensional Fourier transform. The hexagonal geometry reflects that of the material and shows that periodicity occurs in various directions. For the spectrum of the disordered image in (d), we obtain a solid circle, because as we move in any direction in the material, we find the same broad distribution of sizes. The symmetry of the Fourier transform indicates that the material is isotropic: it has no preferred direction.

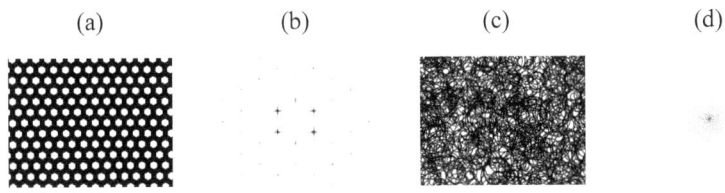

Figure 8.6 (a) Ordered two-dimensional structure. (b) Two-dimensional Fourier transform of (a), which has a geometrical character. The points indicate that periodic structure exists in various directions within (a). (c) Random two-dimensional structure. (d) Two-dimensional Fourier transform of (c), which has no structure. The circular symmetry tells us that all directions are similarly disordered.

If the length scales apparent in (a) and (b) correspond to that of visible light, we would expect a material with such microstructure to favor particular colors that benefit from constructive interference. That is, it should exhibit structural colors due to *coherent* scattering, which does not occur in the highly disordered scenarios that yield white, where no length scale—and hence no color—is favored.

The application of the Fourier transform to investigate the structure of feathers is relatively new, and it resolves a long history of erroneous conclusions about the origins of blue coloration.[8] Even modern references propagate old errors, however, such as the 2017 edition of the classic *Optics* textbook by Hecht, which states that the blue of a jay results from Rayleigh scattering.[9] This idea has persisted, despite work in the 1930s by C.V. Raman, the great Indian Nobel laureate in physics who pioneered the modern understanding of scattering.[10] Raman showed that feathers of an Indian Roller *Coracias benghalensis* could not produce blue by the same mechanism as the atmosphere, suggesting that a coherent mechanism was at work. When electron microscopy was developed a decade later and applied to plumage, however, the structures *appeared* random, and the coherent scattering explanation was abandoned.[11] Then Yale biologist Richard Prum applied the Fourier transform, and the evidence for quasi-ordered structure and coherent scattering became overwhelming.[12] Various studies of blue feathers have now shown that there isn't just a single way to develop this structural color. It may be due to a semi-ordered collection of spherical voids, cylindrical structures, channels or other geometries that conspire to produce reflecting surfaces that have relatively consistent spacings.[13] Figure 8.7 compares the structure of blue and white feathers in terms of their electron micrographs in diagrams (a) and (c), respectively; and the Fourier transforms of these images, in (b) and (d), respectively. The Fourier results make clear that the blue results from coherent scattering and the white from incoherent scattering.

The idea of 'quasi-order' implies that a particular spacing is prevalent, but enough variation exists such that the structure may lack an exact or even

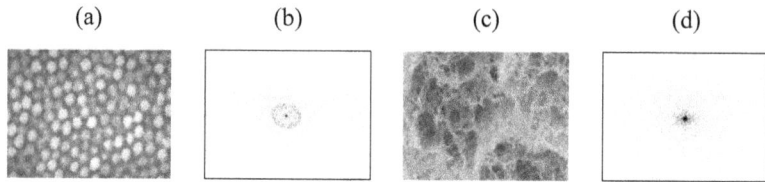

Figure 8.7 (a) Electron micrograph showing quasi-ordered structure in the barb of a crown feather from a Blue-crowned Manakin *Lepidothrix coronata*. (b) Two-dimensional Fourier transform of the image in (a), revealing the clear ring-like signature that accords with coherent scattering. (c) Electron micrograph of a white barb, from the tail feather of a Eurasian Woodcock *Scolopax rusticola*. (d) Two-dimensional Fourier transform of the image in (c), revealing no ring-like signature. Only incoherent scattering can result. [Diagram (a) reprinted from Saranathan, Forster, Noh *et al.* (2012) with permission.[14] Diagram (c) reprinted from Dunning, Anvay, D'Alba *et al.* (2023), under the terms of the Creative Commons Attribution License.]

obvious regularity. This is in contrast to a *highly ordered* geometry, in which the spacings between scattering surfaces have a crystalline precision. Such structures occur often in nature—even in certain bird feathers—where the result is *iridescence*: an effect in which the perceived color changes depending on the orientations relative to the light source, the object and the viewer. In some ways, this is a simpler optical process than what we have studied so far. We turn to this topic next.

Order and iridescence

If quasi-order can give the vibrant color of a bluebird, what might result from 'full' order? We don't mean rigid order at the nanometer scale of atomic spacings. Liquid water has no order, while a diamond is defined by its crystalline regularity; yet both reflect and transmit light in a similar way, because light is so much larger than the interatomic distances. Rather, we mean an orderly arrangement of scattering surfaces, which result from two (or more) materials with different refractive indices. The simplest way to achieve a kind of perfect order is to have two materials kept separate, with a flat plane as their shared boundary. We considered this previously with the reflection and refraction on the surface of a motionless body of water.

This is a bit too simple, though, because reflecting light from one flat surface won't give rise to any interference effects. Something with two parallel flat surfaces, like a pane of glass, fits the bill. Look carefully out through a window at night and you'll notice that well-lit objects inside are reflected twice—where the indoor air meets the glass, and again where the glass meets the outdoor air—because the refractive index changes at both boundaries.

A similar geometry defines a thin film, as shown in Figure 8.8. Some incoming light from the left reflects from the top, while most enters and is

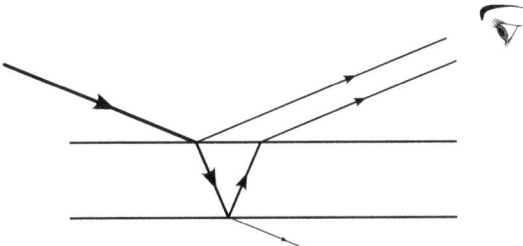

Figure 8.8 Light from the upper left is incident on a thin film. Some light is reflected, while most is transmitted. Some of this transmitted light will be reflected from the bottom surface and re-emerge, parallel to the reflected light. Each of these paths involves the light traveling a different distance before it is seen by an observer on the right.

refracted. The second surface reflects some of this back up, and the reverse transition of the top boundary 'un-bends' the light. An observer will see superposed light coming off both surfaces, with the two rays having traveled different total distances. The different path lengths change the relative phase, leading to constructive or destructive interference, or something in between.

To trace through this carefully, we must note some other details concerning reflection. Whenever a wave meets a 'hard' boundary, the reflection is *inverted*. For example, when the crest of a water wave hits a vertical wall, the rising water pushes upward against it, so the wall reacts with an equal and opposite force down, producing a reflection in the form of a trough in the water, moving back outward.

Something similar happens to a pulse moving down a lightweight rope (along which it travels quickly) that is tied to a heavier one (along which waves move more slowly), as shown in diagrams (a)–(c) of Figure 8.9. At the boundary, some energy is transmitted and some reflected; but because of the reaction force at the boundary, the reflected wave is inverted: it is 'flipped

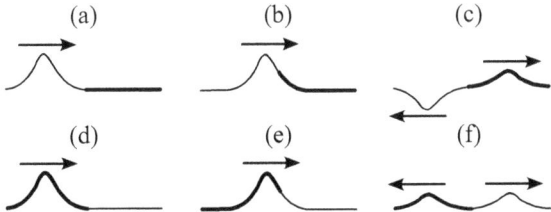

Figure 8.9 (a) A pulse moves down a lighter string from the left toward a boundary with a heavier string. (b) The pulse reaches the boundary. (c) Some of the energy is transferred to a lower amplitude pulse in the heavy string, while a reflected, inverted (upside-down) pulse moves back along the lighter string. (d) A pulse moves down a heavier string from the left towards a boundary with a lighter string. (e) The pulse reaches the boundary. (f) Some of the energy is transferred to a lower-amplitude pulse in the lighter string, while a non-inverted pulse is reflected back up the heavier string.

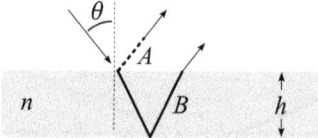

Figure 8.10 Light arrives at an incident angle θ on a thin film of thickness h and index n. Some light is immediately reflected and the rest enters, where some is reflected at the bottom surface and eventually reappears above the film. The different paths are denoted by A (dashed line) and B (solid, heavy line). The difference between $2B$ and A will determine the character of the interference that results.

upside-down'. But if we reverse the situation and have the pulse move from a heavier, slower string to a lighter, faster one, as shown in (d)–(f), the reflection *will not* be inverted. These are universal properties of waves, and these diagrams also represent what happens when light crosses boundaries where its speed is changed.

Now we'll return to the film geometry, as shown in greater detail in Figure 8.10. For simplicity, the surrounding medium is air with $n = 1$. The film has a thickness h and index n, and the light strikes at an angle θ. Our concern is to track the different paths that the light must take before being observed. These differences are shown with the thick lines. The light from the first reflection (dashed line) travels some distance in air, denoted by A, while the light reflecting off the bottom (solid line) travels within the film a distance $2B$. These lengths will depend on θ, h and n.

Looking straight down, we'd have $\theta = 0$, $A = 0$ and $B = h$, so the paths would differ by $2h$. For light of wavelength λ, we will get constructive interference if $h = \lambda/4$. The path difference becomes $\lambda/2$, but the first reflection was inverted, which is the same as a half wavelength shift. A quarter-wave thickness thus has no net phase shift, and the reflections superpose perfectly.

Suppose we set the thickness to 180 nanometers, which is one-fourth the wavelength of red light. Incident red light will be strongly reflected; but other colors will not, because their wavelengths do not accord with the film thickness. They would require thinner films. Hence, we can illuminate it with white light and the perpendicular reflection will be reddish.

If the angle θ changes, so will the difference in the path length; it will always become smaller, so the interference effect must favor shorter wavelengths. A plot of the peak wavelength versus angle is shown in Figure 8.11, for $h = 180$ nanometers and $n = 1.5$. The color shifts from red toward blue, before it disappears (to our eyes) when it moves into the UV regime.

This shift in color versus viewing angle is a signature of iridescence, and it always points to a thin film geometry at work. And we can learn something

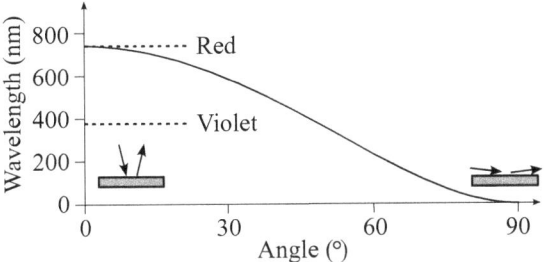

Figure 8.11 Wavelength in nanometers yielding constructive interference for a thin film versus viewing angle for a film of thickness 180 nanometers and an index of 1.5. When illuminated and viewed from above (an angle of 0°), the film will appear red. As the angle increases it will shift toward blue, but it will eventually yield no visible reflection beyond 50° or so.

very interesting about structural colors if we apply this learning to a more general scenario. Suppose we had a collection of thin slivers or platelets that each act as a thin film, but are arranged so as not to be perfectly aligned; instead, they are 'quasi-ordered', as shown in Figure 8.12. We assume they all have the same thickness, corresponding to a quarter of some wavelength. What will an observer see when white light is reflected from them?

Let's first consider the case that these films are very thin: a quarter of the wavelength of blue light. In that case, the platelets that happen to be 'head-on' to the observer ($\theta = 0$) will appear bluish, while those that are tilted will strongly reflect shorter, UV waves that cannot be seen. The result will be a 'pure' blue appearance. (Since birds can see UV, they won't see blue per se, but 'bluish-UV', which we cannot imagine!) But if the film thicknesses are larger, corresponding to a quarter wavelength of red light, we cannot get a purely red result. For, although those reflections at $\theta = 0$ will be red, those coming from the tilted platelets, with $\theta > 0$, will reflect shorter wavelengths that we *can* see—corresponding to oranges, yellows, greens and blues—leaving us with no color at all. This is why there is quasi-ordered, structural blue, but not red. *Structural red* can only be produced by highly ordered thin films, and its brilliance will fade as the angle changes. And so, for example, the red gorget

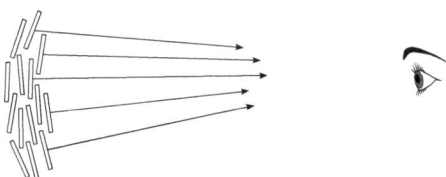

Figure 8.12 A quasi-ordered material, in which a variety of reflection angles are obtained simultaneously. This will result in a mix of wavelengths equal to or shorter than the nominal wavelength for constructive interference.

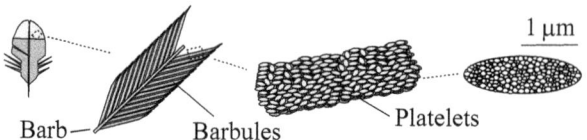

Figure 8.13 Structure of a hummingbird gorget feather on the left. Only the distal portions (shown as a lighter shade) produce iridescence. A close-up shows a single barb and many of the barbules arranged in a V-shape about it. Close inspection of a barbule reveals multiple layers, which each contain platelets that produce the iridescence. A close-up of a single platelet shows it is several microns long and contains a mix of air pockets. [Modified from Greenewalt (1991).]

of a hummingbird is best seen head-on, losing its hue at an oblique angle, eventually turning black.

The simple film geometry in Figure 8.10 permits much of the incident light to 'escape' via the lower surface. Reflection off a transparent material typically involves only a few percent of the incident radiation, so interference from a single film cannot produce copious light. But if we stack multiple films atop one another, we can produce more reflections. Moreover, we are not limited to two materials; we could add layers with other refractive indices, and with some care as to their thicknesses, reinforce the interference and produce a more brilliant effect.

When we look for the origin of the hummingbird's iridescence, such stacks of reflecting layers are exactly what we find, localized to the tiny barbules.[15] Figure 8.13 shows the details of a gorget barbule for an Anna's Hummingbird *Calypte anna*, which presents planar surfaces densely packed with thin elliptical platelets. These form a mosaic when viewed from above, and are found to form stacks of parallel layers when viewed from the side.

Barbules are primarily keratin, and the platelets can include large air pockets. But the platelets also contain melanin—a pigment responsible for blacks, browns and grays—and we will look at this in more detail shortly. A black pigment may seem paradoxical in a brilliant feather, but it occurs in very thin layers that absorb relatively little light, has a very high index of refraction (1.7 to 2.0) and reflects well.

The oval platelets built of melanin are called melanosomes, which may be filled with melanin, air or a mixture, and may be cylindrical in shape in the iridescent feathers of other species. Trogons, starlings, pheasants and sunbirds are just a few of the families that employ iridescence. The geometrical details differ but the underlying mechanism is the same.[16]

Many hummingbirds display striking green iridescence, which can be dialed in via the appropriate platelet details. Some green feathers, notably, can exploit a similar structure and yet lose the iridescent effect, making their vibrant color visible from all directions. The trick here is that the platelets do not lie flat, but rather form a concave contour, so that regardless of the viewing angle, some portion of its surface faces the viewer—producing the green—while the rest of the surface contributes only shorter wavelengths: blues that don't too appreciably shift the overall verdant hue.[17] The situation is similar to that in Figure 8.12, but with the platelets arranged to form a 'C' shape.

The green feathers of most birds, however, rely on the quasi-ordered structures of the previous section to produce blue, but augmented with yellow produced by pigmentation. A notable exception are turaco feathers, which have no structural color, but rely on a novel green pigment specific to this bird family.[18] Pigmentation generates color at the molecular level; it doesn't matter how the molecules are arranged. Grind up a blue feather finely enough and it will produce a colorless substance, but a red feather will leave a pile of red powder. We'll now switch gears to consider this utterly different method of making color.

Reds and blacks from pigments

The mechanisms that we have focused on thus far have involved scattering; light jostles the bound electrons in some material and they re-radiate it, with the interference effects dictating the results. But matter and electromagnetic radiation can also interact via the process of absorption, in which molecules will 'soak up' particular wavelengths, but not others. When exposed to incident white light, portions of the spectrum then effectively disappear into a material, while the rest is reflected back with some residual color. An organic molecule that readily absorbs some amount of visible light is a good working definition of pigment.

In feathers, pigments are typically either melanins—which produce black and brown colors—or carotenoids, which are responsible for yellow, orange and red hues. The two forms are quite dissimilar in terms of their arrangements of atoms, but their optical properties arise from similar physics. These molecules are large and complicated, and to better understand how they interact with light, we'll start with the simpler properties of their constituent atoms.

For an atom to absorb light, it must convert the incoming energy into its own excess energy by moving an electron into an excited, unoccupied energy state. Since only specific energies are allowed by the various atomic orbitals, the incident light, which comes in discrete packets—photons—must carry a precise amount of energy to be absorbed. That energy is inversely proportional to wavelength; so that a red photon carries more energy than an infrared photon, a blue photon even more, a UV photon more still and so on. Because

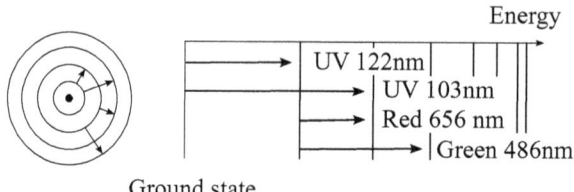

Ground state

Figure 8.14 The left side is a schematic (not to scale) of the energy levels for a hydrogen atom, with arrows indicating transitions between various states when light is absorbed. The right side shows the energy levels as vertical lines. Various transitions between them are marked in terms of the light wavelength needed for absorption to occur.

there is a fixed set of orbitals, there is a limited set of wavelengths that can be absorbed. Some energy levels and wavelengths for the simplest element, hydrogen, are shown in Figure 8.14. Every element has a unique set of energies and therefore a distinct set of spectral lines—so-called because a display of the resulting colors after light has passed through a sample will be missing the absorbed wavelengths, which appear as black lines.

The same idea applies to molecules, in which two or more atoms achieve a stable configuration by sharing unpaired electrons to form bonds. In the case of the N_2 molecule, the electrons are tightly constrained and can only absorb energetic, short-wavelength radiation in the UV range.

For larger molecules, the situation becomes more complicated, as the electrons are pulled on by many atomic nuclei; this leads to a more complex collection of *molecular orbitals* that each has its own specific energy level. An energy diagram similar to that in Figure 8.14 will still be obtained, but it will have many more levels. Some levels will be closer together, permitting longer wavelengths to be absorbed because less energy is needed to bump an electron up. We will estimate the effect of molecule length on absorption frequency shortly.

The absorption of light by molecules is made even more complicated by the dynamic nature of the molecule's shape. The nuclei comprising a molecule are bound together, but not rigidly fixed in place; we might picture them as billiard balls connected by springs. Being relatively massive, the frequencies at which they vibrate will be relatively low, typically in the range of infrared radiation. This atomic motion affects the energy levels of the electrons, so with more ways for a molecule to vibrate, twist and otherwise subtly change in shape, the number of specific wavelengths of radiation that can be absorbed will grow. The upshot is that larger molecules develop finely spaced sets of energy levels, and instead of the thin lines of spectral absorption found with atoms, they begin to exhibit 'bands' over which they can soak up radiation.

Carbon atoms are well suited for building large molecules, as they have four valence electrons that are eager to form bonds. Starting with a few carbon and

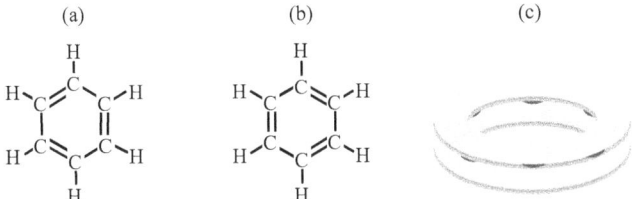

Figure 8.15 (a) The benzene molecule, consisting of a ring of six carbon (C) atoms which each bond to their two C neighbors and a single hydrogen (H) atom. The sets of two lines indicate double bonds. (b) The same molecule, but with an alternate configuration of bonds; together with (a), this comprises a resonant structure. (c) A depiction of the π orbitals, indicating that electrons can freely move around the molecule.

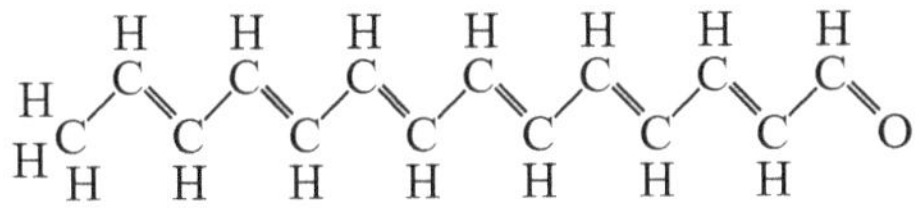

Figure 8.16 A tetradecahexenal molecule, primarily consisting of hydrogen atoms (H) bonded around a long chain of carbon (C) atoms connected with conjugated (alternating single and double) bonds and a single oxygen atom (O).

a handful of other atoms, we will quickly find that there are many ways they can be combined. An important example is the benzene molecule, which has six carbon and six hydrogen atoms arranged in a flat ring-like structure, as shown in Figure 8.15 (a) and (b).

The lines indicate covalent bonds, which consist of an electron pair. Every hydrogen atom has one electron, which ties up one from its carbon neighbor, leaving the latter three valence electrons each. These also look to bond, but there are only two adjacent carbons. This is dealt with via double bonds, indicated by two parallel lines, in which each carbon shares a pair of electrons.

Note that there are two equivalent ways, shown in diagrams (a) and (b), to configure these bonds; but neither is *preferred*. Nature deals with this by allowing both: a superposition, if you will. These states form a *resonance*, with the upshot that the electrons in the resulting orbitals (denoted π orbitals) become delocalized and free to move around in rings, as shown in Figure 8.15 (c).

The alternating of single and double bonds is referred to as *conjugation*. In this example, the geometry is cyclical, but linear molecules can feature conjugation as well. Figure 8.16 shows a single molecule of tetradecahexenal: a pigment used exclusively (among birds) by psittacids such as parrots.[19] Because of the alternating double and single bonds, electrons can effectively roam the length of the molecule. In a sense, it is almost like a little wire antenna, supporting an electrical current to flow up and down.

Electrons are waves and, because they are confined to the length of the molecule, their wave amplitude must be zero at the endpoints; just as a guitar string cannot vibrate at the ends where it is clamped down. Specifically, if the electron is restricted to a region of length L, the wavelength λ is limited to values such as $2L$, L, $2L/3$ and so on; in other words, $\lambda = 2L/N$ where $N = 1,2,3$ etc. The situation is entirely analogous to the resonant wavelengths of sound in a tube, as shown in Figure 5.7, but with zero amplitude at both endpoints. This has huge implications for how light will affect a molecule in which an electron can roam a length L. To see why, we will create a simple model of this scenario.

We begin by spelling out two foundational rules of quantum mechanics, discovered in the early 1900s. Neither of these laws can be derived from other principles; they are simply a matter of observation. The first states that, although light is an electromagnetic wave, it yet comes in the form of discrete particles, each with energy proportional to frequency, according to $E = hf$, where h is Planck's constant. The second is that a particle (such as an electron) that has mass and momentum is yet a wave; and the shorter the wavelength, the larger the momentum p, according to $p = h/\lambda$. And momentum, going back to Newtonian physics, is related to kinetic energy according to $E = p^2/2m$, where m is the mass. In terms of wavelength, this becomes $E = h^2/2m\lambda^2$. The allowed energies for the electron that is confined to a length L are then given by:

$$E = \frac{h^2 N^2}{8mL^2}$$

When light is absorbed, the only thing that changes is N, going from one integer (N_0) to a larger one (N_1). Since the energy of the incoming light must be equal to the change in electron energies, and the light's frequency is given by $f = E/h$, we have:

$$f = \frac{h}{8mL^2}(N_1^2 - N_0^2)$$

All other things being equal, the larger L of a long molecule means that less energy is needed to excite the electron, and so a lower frequency of radiation can be absorbed. This moves the absorption into the range of visible light and is the key to color production via pigmentation.

Experimentally, it is found that at least seven conjugated pairs are needed for absorption to occur in the range of visible light; fewer than this and we'll only have UV absorption, and so no color is produced.[20] With seven conjugated pairs, blues are primarily absorbed, leaving the rest of the visible spectrum to take on a yellowish cast. Add more conjugated bonds and even lower wavelengths get absorbed, leaving a progressively redder character to the reflected light. Hence, a range of colors can be 'tuned in' according to the molecule's size.

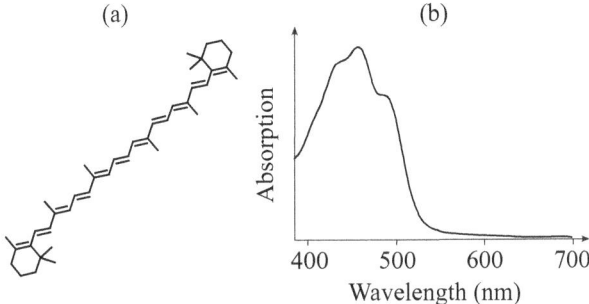

Figure 8.17 (a) The beta carotene molecule. At every junction sits a carbon atom (not labeled). (b) Corresponding absorption spectrum, which shows a large peak for shorter wavelengths. Such a molecule will produce a reddish color because long wavelengths are not absorbed. [Spectral data taken from Johnson (2016).[21]]

Except for parrots, birds produce reds and yellows from *carotenoid* pigments. At least 34 different types have been found in feathers; and unlike the parrot's psittocoflavins—which simply feature different lengths—these carotenoids employ the same linear chain but differ at their ends, where benzene-like ring structures may cap them off.[22] These effectively change the 'electrical length' of the molecule and hence tune the color produced. A model of a typical carotenoid (beta carotene) is shown in Figure 8.17, along with its visible light spectrum, showing that it readily absorbs short wavelength (violet and blue) radiation.

Such molecules are not trivial, but their optical properties and their chemical makeup are generally known. The same cannot be said for the melanins, a family of black and brown pigments whose structural details and behaviors are still not fully understood.[23] The very property that makes them so novel and useful—their broad absorption of the visible and UV light—frustrates the same spectroscopic tools used to analyze other molecules.[24]

Melanins are nothing like the linear carotenoids; they are more akin to two-dimensional mosaics formed by benzene-like rings. Figure 8.18 (a) shows a portion of such a structure. Diagram (b) shows a smooth and broad absorption profile typical for melanins.

It is not difficult to see how this circuitous maze results in long sequences of conjugated bonds, and subsequent absorption of visible light. It has a very smooth broadband character believed to arise from disorder in these complex structures. Chemical disorder leads to a range of absorption peaks, while geometrical disorder among neighboring molecules—which may stack up in close planes—contributes additional ways for the energy levels to multiply and effectively smear out, creating a kind of net that readily captures any wavelength thrown at it.

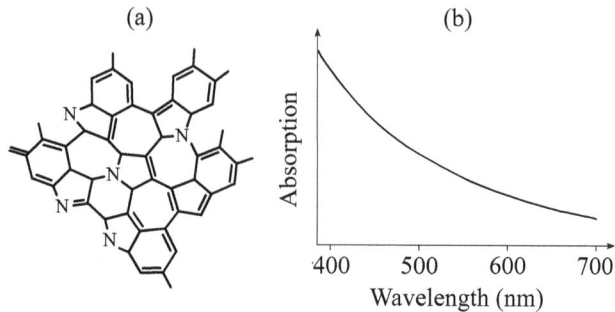

Figure 8.18 (a) Portion of the structure comprising a melanin molecule. Carbon atoms (not labeled) reside at each bond junction, except for locations occupied by nitrogen (N) atoms. (b) The corresponding absorption spectrum covers the entire visible range. [Diagram (a) redrawn from Costa, Feldhaus, Vilhena et al. (2014).[25] Spectral data in (b) taken from Ahmad, Alqahtani, Al-Terary et al. (2019).[26]]

A molecule that absorbs light enters an excited state with higher energy, but must eventually do something with it. With atoms, what goes up comes down in the form of emission; sometimes equivalent to the absorbed light, sometimes as multiple photons, as the atom steps down through the available energy levels. With large molecules, the energy gets coupled into the movements of the atoms within the structure, and is eventually released as heat.

While absorbed light is typically coupled into vibrations, some pigments leverage the energy to undergo structural changes, which can be exploited to make a light-sensing transducer. This is why pigments, besides helping to create colors for us to perceive, also make the physiology of vision possible; we will take this up in the next chapter.

Chapter 9

In the Blind: Images, Eyes and Cameras

Staking out shy birds from the comfort of a blind, we reflect on the 'back half' of enjoying a good view: our visual perception. Without a device to form an image, all that color would amount to nothing. In this chapter, we'll examine how eyes—as well as their human-made counterparts, cameras—operate. We'll see how everything involves performance trade-offs, leading to a plethora of photographic equipment and remarkable strategies that birds use to achieve extreme visual capabilities.

Natural and human engineering

Before we were here to appreciate the colors of birds, and before there were birds, light and matter followed the same rules of engagement. The hottest stars were radiating primarily blue light, and the cooler ones red. Atmospheric gases split incident light via scattering to produce multicolored twilight skies. Crystalline structures of various minerals possessed the right length scales to cause brilliant, iridescent reflections. Even pigments were at work, with the ubiquitous chlorophyll molecule making plants reflect green light although there were no eyes to appreciate it.[1]

We've seen how pigments utilize conjugated bonds to absorb radiation and produce—among other things—red, yellow and black feather colors. But pigments are also a critical component within the light-sensitive cells of our eyes and have as much to do with our sensing of light as with the generation of color. To understand how they make vision possible, we need to continue the study of how matter absorbs radiation and ask where the energy goes after an electron has acquired it.

Typically, a molecular electron that has absorbed light will lose that energy due to its coupling to nearby nuclei. This can manifest as vibrational and rotational motion of the molecule as a whole, which raises the temperature and ultimately re-radiates light, now as infrared radiation.

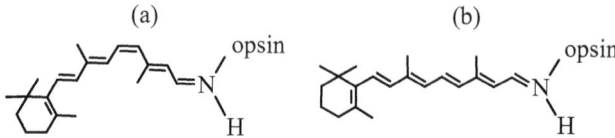

Figure 9.1 (a) The cis isomer of the retinal molecule attached to an opsin protein. All bond intersections hold carbon atoms (not illustrated). (b) With the absorption of light, the retinal molecule isomerizes to the all-trans form, leading to a cascade of changes in the opsin protein.

But the energy can also cause *deformation*. In certain molecules, electron excitation forces the molecule to *isomerize*: that is, to change shape without changing its chemical makeup. Such is the case for the pigment *retinal*, a molecule in the rod and cone cells that cover the eye's retina.[2] The two states of the pigment are shown in Figure 9.1. The lower energy configuration in (a) can be altered to the isomer in (b) via the absorption of light.

In the cell, retinal is attached to an *opsin* protein, and the light-induced shape shift leverages a change in *its* form, kicking off a chain of molecular reactions. What follows is too complex to trace through here: a change to the flow of ions across the cell membrane, triggering the electrical signals, the release of neurotransmitters, and eventually the signals in the brain that generate the conceptualization of the image.

The isomerization of retinal by light, and the many subsequent events, are just one part of the process. A great deal of biochemistry must run in the background to enable the sequence to occur repeatedly, for after a retinal molecule has been altered, we are left with something like a sprung mousetrap that needs to be reset.

All of this takes place at one end (the 'outer segment') of the rod and cone cells in the retina. Both types of cell feature stacks of disk-like membranes full of opsins, thereby increasing the sensitivity to light. Diagrams of these cells are shown in Figure 9.2. In vertebrates, these cells face the 'wrong way' so that light not only needs to traverse the long axis of the cell, but the 'wiring' must be routed out via the blind spot.[3]

The rod cells are found in the periphery of the retina and provide low-light vision but no color discrimination. The cone cells, which provide central vision, come in three varieties, each sensitive to a different peak wavelength. By discriminating the relative levels of three signals—roughly tuned to red, green and blue—we can recognize more than two million distinct colors. The spectral tuning of a given cone cell is determined by the details of how the retinal is attached to the opsin. The bond gives the retinal the leverage to alter the protein, but also affects the effective number of conjugated bonds and hence the peak wavelength.[4]

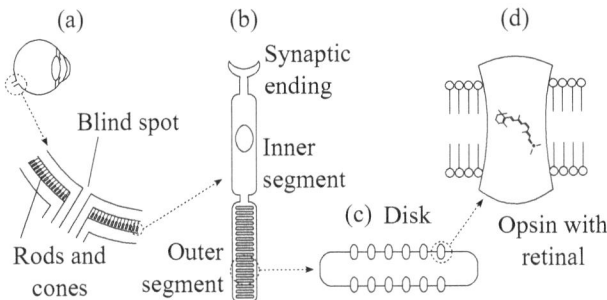

Figure 9.2 (a) Schematic of the eye, highlighting a portion of the back of the retina, including the blind spot where the optic nerve (in gray) passes out. The rods and cones sit beneath the layer of nerves and are directed away from the incoming light. (b) A simplified rod cell, with a synaptic ending that will initiate a nerve impulse; an inner segment holding the nucleus; and an outer segment with a series of disks. (c) Simplified close-up of a disk, which contains a number of light-sensitive opsin proteins. (d) Schematic of the opsin protein holding a retinal molecule, which will absorb light and lead to structural changes. [Diagram (a) modified from Marieb and Keller (2021).[5] Diagrams (b)–(d) modified from de Grip and Ganapathy (2022).[6]]

An individual cone cell is slender: roughly 50 microns long and 1 to 4 microns in diameter. It presents a tiny cross-sectional area capable of sensing light. That the area is so small is critical for the formation of a detailed image, which we'll consider shortly. But first, we'll look at the visual sensors that *we* have engineered, specifically those found in a modern digital camera.

Carbon has no rival for the title of Most Useful Element, at least for biology. But the element below it in the periodic table, silicon, wins that award in the context of technology. Both elements gain utility from having four valence (outermost) electrons, but the usefulness of silicon derives from how we can manipulate its proclivity to either conduct electricity or not: hence its designation as a *semiconductor*.

Every semiconductor device relies upon junctions between two materials that have different electronic properties. The two materials are both primarily silicon, but they differ critically in that they each have been *doped* with different 'impurity' elements. In pure silicon at room temperature, each atom uses its four valence electrons to bond with its neighbors, leaving it generally unwilling to conduct electricity. If we introduce boron as a dopant into silicon, such that it accounts for about 1% of the total, the conductivity is much better. This is paradoxically because boron has three valence electrons, so that it cannot bind with all of its silicon neighbors, leaving a 'hole' in the electron structure. Such a hole can hop from one atom to another, effectively enabling current to flow.

More obvious is the outcome of doping silicon with material that has five valence electrons, such as phosphorus. This leaves electrons that cannot participate in bonds free to roam about, and hence support the flow of current.

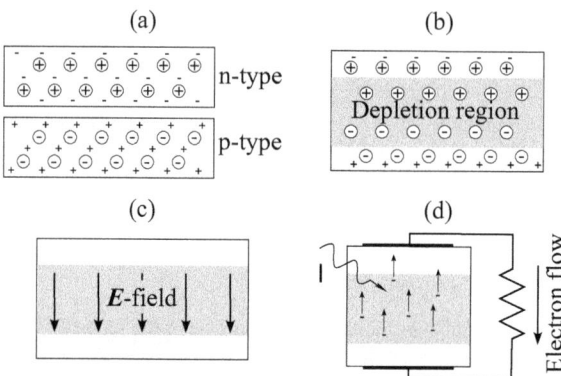

Figure 9.3 (a) An n-type semiconductor contains excess negative charge carriers (–) as well as excess positively charged dopant atoms dotting the silicone lattice. A p-type semiconductor has excess positive charge carriers and negatively charged dopant atoms. (b) Bringing these two materials in contact will allow the charge carriers to flow until a stable configuration is reached, resulting in a depletion region with no net surplus of positive or negative carriers. (c) The dopant atoms, fixed in the lattice, create a non-zero electric field. (d) Incident light arrives in the depletion region and is absorbed by electrons, freeing them from their local bonds and causing them to move along the electric field lines, which will lead to a current flow if the two layers are connected across a resistive load.

Such a material is said to be an *n-type* semiconductor because it has negative charge carriers, while the boron-doped version is of the *p-type*, for positive.

By placing n- and p-type semiconductors in contact, we can create various clever gadgets, such as a *photodiode*, an electrical device that produces a current when exposed to light. A simplified version is shown in Figure 9.3. Starting with n- and p-type layers, as in (a), an electrical junction is formed by placing them in contact (b). Excess electrons above the junction now have a place to go—the holes of the bottom layer—and they will re-distribute themselves. But there is a limit to how many can cross the junction. We must keep in mind that the dopant nuclei in each material (shown by circles) are part of a solid and not free to move. After some number of electrons have equalized holes, we are left with more positive charge above (and less below) the junction, due to the dopant nuclei, and this causes an electric field, which opposes the flow of any more electrons from the n-type layer (c). There is now a narrow region that cannot be crossed *in this direction*: a kind of demilitarized zone called the *depletion region*.

Now suppose light penetrates this structure and is absorbed by an electron in the depletion region, knocking it out of its bond. If this occurred in a block of ordinary silicon, the electron would wander randomly about—perhaps eventually popping back into place—but in the depletion region of our junction, the electric field will act to force this electron up into the n-region. If we had connected wires to the two sides of this junction (d), the electron

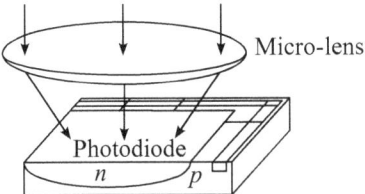

Figure 9.4 A simplified camera sensor consists of a micro-lens that focuses light onto a photodiode, which is simply a junction of n- and p-type semiconductors. Additional circuitry is fabricated onto the semiconductor as well.

would contribute to a current that would flow out and through some external circuit. This is the basic idea of transducing light into electrical energy, which is exploited by solar cells and camera sensors.

Just as our biological sensors require much more than a retinal molecule changing shape when struck by light, so too is the entire operation of a photodiode pixel more complicated than what we have shown. This photodiode does not run continually. When light strikes, the electrons are collected for some time before they are siphoned off and used to control a transistor—another device made from n- and p-type materials, which acts as an amplifier, in a kind of electrical cascade. After this result is read back, the circuit is flushed of any excess charge in preparation for the next exposure.

A typical digital camera may use photodiodes with a cross-sectional area ranging from several square microns to tens of square microns.[7] In other words, their area may be comparable to that of rods and cones. However, the area is not entirely used to collect light; it is partly taken up by the electronics needed to amplify and manage the signal. A micro-lens is positioned above each sensor in order to mitigate this loss of space. A simplified version of a typical sensor is shown in Figure 9.4.

As is the case with biological vision, a single digital sensor will not provide color information. We must resort to a similar trick: using three sensors, each tuned to red, green or blue. This is done by placing filters over each element. By comparing the current produced by three different sensors, we have the similar capability to resolve millions of colors. This alone is a great feat, but in order to go the next step—to the generation of images—we'll need to work with many more sensors incorporated into an array.

Sensors and images

We have seen two different ways to make a light-sensitive transducer, converting radiant energy to electrical or electrochemical energy. Whether based on carbon or silicon, these devices produce a signal proportional to the

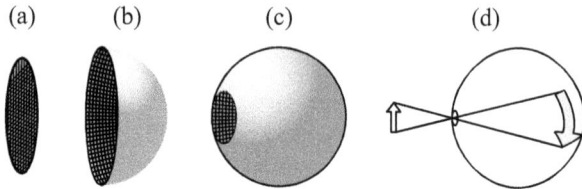

Figure 9.5 (a) A flat array of sensors. (b) Adding curvature to the array allows different sensors to be pointed in different directions simultaneously. (c) A spherical array of sensors might seem counterproductive, as most of the light is blocked. (d) The utility of a small aperture and an array of sensors is apparent when tracing light rays from an object, which reveals that an image is formed.

light detected and are amenable to further 'tuning' to sense only particular regions of the spectrum. This means that, with three sensors dialed in to long, medium and short wavelengths, we can analyze the signal levels and discern the light's color. This is a good example of the utility of having multiple signals to compare. But there is another way, with great practical benefit, in which the signals from many sensors can be exploited. For simplicity, we'll develop this idea based on monochromatic transducers that pick up light and produce a proportional signal; we won't worry about color for the moment.

A single sensor can be useful because it tells us something about the environment. It will alert us if light is present, and perhaps give some clue as to its direction. If the sensor presents a flat, limited area that light can fall upon—as is the case with retinal disks and photodiodes—the energy it receives will depend on the angle at which the light strikes the surface. The difference between the intensity of the overhead sun versus that which comes at a more glancing angle during the late afternoon is a palpable example of this.

By orienting the sensor to maximize the signal, we can eventually determine where the light is coming from. But had we another sensor, always oriented in a different direction from the first, we'd start with more information; the one giving the larger signal would be better aligned with the source. Adding more sensors that point in yet other directions, and then noting their relative output levels, would provide even more detail.

One way to manage this growing array of sensors is to place them in flat sheet and gradually change its shape. If we progressively make it into a bowl, we will increase the variety of sensor orientations in an orderly way, as shown in Figure 9.5. Comparing diagrams (a) and (b), it is clear that the latter will immediately give us more information about the incoming light. Something like this hemispherical bowl might seem optimal because if we continue this trend, we will end up with the nearly closed sphere of diagram (c), which seems to defeat the purpose—the very shape is blocking most of the external world we are trying to see. But in leaving only a small aperture for the light to enter, something momentous occurs: we produce an *image*, as shown in (d).

An image is a mapping of a three-dimensional scene onto two dimensions: a kind of projection, which keeps certain spatial relationships intact. We are so accustomed to being confronted by images of the external world, via two eyes and a lot of signal processing in our heads, that it seems odd to consider what they actually are.

Note how in diagram (d) a scene consists of various sources (here, the different parts of an arrow) that each radiate light in every direction, with only a small fraction passing through the hole. The presence of the hole means that only one sensor (or a few of them in close proximity) will be exposed to a ray from each source, while being shielded from the others. The blocking of light is just as important as the admittance of it.

We note that the spherical shape of the array isn't necessary for this effect; we might take a rectangular box with a flat array of sensors on one side and small hole on the opposite one, and an image will be produced. This is a *camera obscura*, a light-tight box or room illuminated by a single small hole, and has been used by artists for thousands of years.[8] The camera only became photographic—a term that means 'writing with light'—in the 19th century, with the development of plates bearing photosensitive chemicals that could make a permanent record of an image.

There is an obvious limitation to this device, however: the small hole creates an image, but a very *dim* one given the paucity of light admitted. We might address this by increasing the aperture, or width. This will indeed brighten the image, but at a cost. Consider the situation from the point-of-view of a sensor. As the aperture increases, light from other parts of the external world that were previously hidden from it (but seen by neighboring sensors) is now visible. The total light is greater, but it is no longer capturing such a selected part of the world. There is a washing out, a blurring of the image: a loss of focus.

We have arrived at the first of many trade-offs; this will be a constant refrain as we look to improve any kind of optical instrument. We'd like to skirt this dilemma, if possible, and get a brighter image, while keeping the focus sharp. This can be done, but it will come at a cost and necessitate yet other trade-offs. What we need is a method to redirect the light passing through a large aperture in specific ways. It isn't hard to draw the paths we'd like light to take, simply by thinking about the requirements. This process is shown in Figure 9.6.

Diagrams (a) and (b) compare the effect of the differently sized apertures by considering how the light from a particular point, the tip of the arrow, enters both devices. In (b), we see that the wide aperture admits light radiating out from one point to illuminate a swath of sensors. What we want is for all this

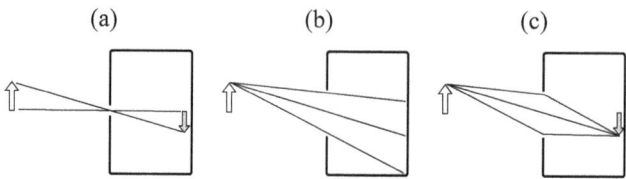

Figure 9.6 (a) An object (white arrow) before a simple camera obscura with a small aperture, producing an image (gray). (b) Opening the aperture admits more light, but the image is lost. (c) In order to obtain an image with a large aperture, the rays from the object would need to be redirected as shown.

light to instead arrive at a single spot. In (c), we draw a set of paths that the light would need to take for this to occur.

To use a wide aperture, we'll need something transparent that can alter the direction of the light. We recall from Chapter 8 that light entering a medium of different refractive index at an oblique angle will be bent. So, as a first attempt at finding a solution, we might try filling up our camera with a clear liquid or glass, as shown in Figure 9.7 (a).

The refractive effects do indeed act to bring the rays closer together on the array, so this seems a step in the right direction, but it isn't a solution. We guess that using less material might be desirable, so let's see what happens if we simply put a plate of glass in the aperture. As diagram (b) shows, the effect is similar. But even if we keep increasing the refractive index n and the angle of refraction, it does not appear that we'll get the sharp focus that the tiny aperture enabled.

We need to get creative. We know that the angle of incidence of the light will affect the amount of refraction as well. Because the light coming through the center is already going where we want it, we'll leave this part alone. Instead, we might introduce some beveled surfaces (as shown in Figure 9.8) along the top and bottom of the glass.

In diagram (a) we see that the light is now going more where we want it. As we try to further bring the rays to a point, we'll find that adding shape to the back side of the glass helps, as does increasing the number of bevels.

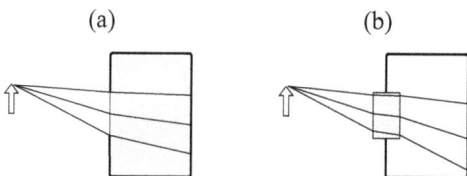

Figure 9.7 (a) Filling a camera obscura with highly refractive material redirects the rays but does not focus them. (b) Using a block of such material in the aperture has a similar effect.

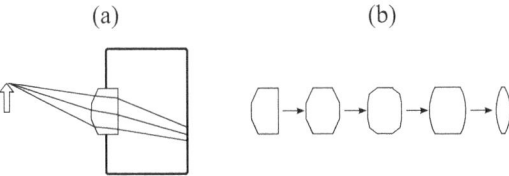

Figure 9.8 (a) Introducing bevels to reshape the glass in the aperture brings the rays closer together. (b) Continuing this strategy in an effort to focus the light, we will eventually arrive at a smooth lens.

Eventually, this progression leads to a smooth, curved shape, as shown in (b). We've just engineered a *lens*, seeing in the process just how it manages to solve our dilemma. In the next section, we'll explore some of the possibilities and problems that our new device affords.

Cameras and trade-offs

A light-tight enclosure with an array of sensors illuminated by a lens can function as an eye or a photographic camera, depending on the details. In broad strokes, they are quite similar, and much that pertains to one applies to the other. Since cameras have become such common birding tools, we'll consider our device from that perspective first. Digital photography is a vast field, full of minutiae that could fill many books.[9] We'll only cover the most basic physics here.

Lenses are typically, though not always, spherical in shape. That is, a biconvex lens is effectively the intersection of two spheres with radii R_1 and R_2; a kind of a 3D Venn diagram, as shown in Figure 9.9 (a). The material has a refractive index n, and outside the lens, we will assume we have air, with an index of 1.

For a sufficiently distant object, the light will reach the lens in parallel rays, as shown in diagram (b). That is, the light striking one part of the lens arrives moving parallel to that at any other part (this is not the case if the object is near the lens, as discussed below). The distance f is the *focal length* and is a critical

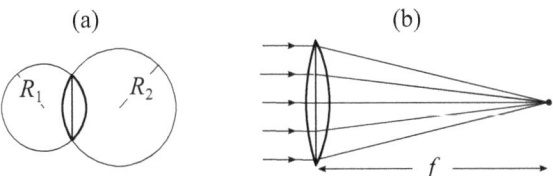

Figure 9.9 (a) A spherical, bi-convex lens has a shape defined by the intersection of two spheres of radii R_1 and R_2. (b) Parallel rays arrive perpendicular to the lens and are brought to a focus due to refraction. The distance from the lens to the focus is the focal length f.

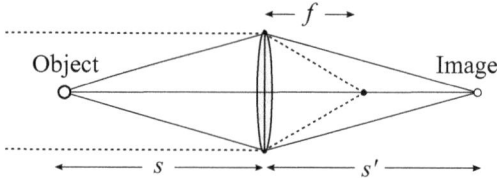

Figure 9.10 An object or light source is a distance *s* to the left of a positive lens. Since the light rays (solid lines) reaching the lens are not parallel as they would be for an infinitely distant source (shown by dashed lines), they will not focus at the focal length. Rather, the light will focus a distance *s′* beyond the lens.

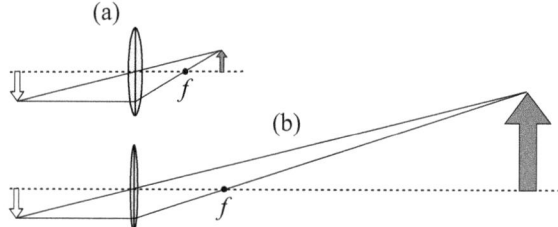

Figure 9.11 (a) A lens with a short focal length produces a small image of a nearby object. (b) Using a long focal length lens, the same object at the same distance will produce a larger image, further back.

parameter for any lens. We can intuit how f is related to various parameters. If R_1 or R_2 decrease, the lens becomes more curved, bringing the focus in closer. If the index n is made larger, the refraction becomes more severe, so we expect f to decrease as n goes up.

Focal length tells us how far from the lens we must place our sensors in order to capture the focused light from a distant source. Using any other distance will produce an image that is out-of-focus: making, for example, a star appear as a circular blur. But if the object is sufficiently nearby, the requirements for focus will change. To see why, suppose we start with a 'point' source of bright light, such as a small LED placed near a lens, as shown in Figure 9.10. Here, geometry dictates that the light cannot reach the lens as parallel rays, as it would for a very distant source. By shifting the angles of incidence, we also change the angles at which the rays exit the lens, moving the focal point out to a distance $s′$, beyond f.

If the sensors remain a distance f from the lens, the image will be unfocused; something will have to change. In the case of our eyes, the response is to alter the lens curvature (using the small muscles that support the lens) and decrease its focal length, which will bring the focus onto the retina. In a camera, we may turn a focusing ring, which increases the distance between the lens and sensors out to $s′$.

Since we can design a lens to have whatever focal length we choose, we need to know the role of f in a camera's performance. Convenience often favors a smaller value, as this keeps our device from becoming large and unwieldy. But a shorter focal length results in a smaller image, as shown in Figure 9.11. We can immediately see why greater magnification comes with long lenses; there is more distance for the image to 'grow'. Some other ways to enhance magnification will be discussed in the next chapter.

Another consideration is the lens aperture, which determines the total amount of light that can be admitted. It might seem that a larger aperture would always be beneficial, and if astronomy is your pursuit, then this is indeed the case; only by harvesting more light can distant, faint objects be perceived. For terrestrial photography, the benefit of a large aperture is that it requires less exposure time: that is, faster shutter speeds. The quicker an image can be captured, the less susceptible it is to motion blur, be it from a fast-moving subject or a shaky hand.

The potential liability of a large aperture is that the *depth of field* (DOF) is sacrificed. DOF refers to the range of distances between objects and lenses that can be adequately focused at the same time. Figure 9.12 (a) illustrates a large-aperture lens of focal length f and two objects, O_1 and O_2, at different distances. They will each create images I_1 and I_2 at different locations behind the lens, and therefore cannot simultaneously be in focus no matter where the sensor is. Here, we have placed the sensor at a midpoint between the images,

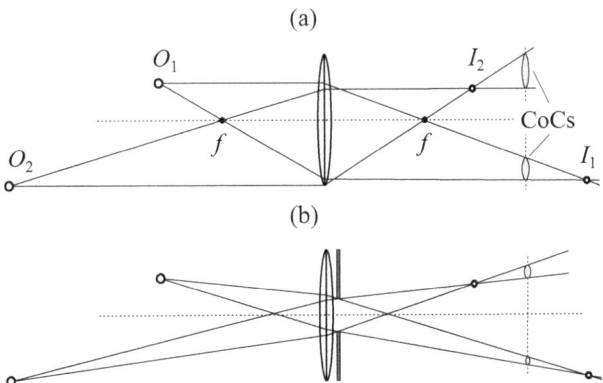

Figure 9.12 (a) Two objects O_1 and O_2 lie at different distances to the left of a lens. Ray tracing shows that they produce images at I_1 and I_2, respectively, and they occur at different distances to the right of the lens. Placing our sensor of film at a position midway between them (dashed vertical line) would produce two spots, each an out-of-focus image. These are the 'circles of confusion' or 'CoCs'. (b) A constricting aperture is placed directly behind the lens, allowing only the central portion to pass light. Reconstructing the ray diagrams shows that the CoCs will be reduced in size.

where it will capture only blurred spots, known as 'circles of confusion' (CoCs). In (b) the aperture is reduced by an iris that passes light only from the central part of the lens. Tracing the course of the rays, we see this must reduce the size of the CoCs. Eventually, such defocused spots can be made small enough that the image has sufficiently sharp detail.

Aperture is easily changed on demand via a variable, iris-like diaphragm that expands or contracts. Since lens diameters vary, it is customary to describe the aperture not in terms of absolute size, but rather as a ratio to focal length, by means of *f-stop*. An f-stop of 'f/4' sets the aperture to be four times smaller than the focal length.

If we desire both good depth of field and a short exposure time, we have a dilemma, because a small aperture and fast shutter speed will conspire to make a dim, or underexposed, image. The only solution is to use sensors that are more responsive to low light levels. Before digital photography, sensitivity was determined by the size distribution of the light-sensitive silver halide grains of the film. A 'fast' film had larger grains (on average) that could more rapidly collect light than the smaller grains of a 'slower' film. Hence, the designation of film 'speed', quantified by ASA (American Standards Association) or ISO (International Organization for Standardization) ratings. The trade-off here was exposure time versus resolution, or graininess. A low ISO meant smaller grains and better resolution, but longer exposures.

With a digital sensor, the term ISO has been maintained as a way to quantify the third side of the 'photography trilemma' (the others being aperture and exposure). But ISO has become more complicated and there is no single definition, as camera vendors will use different methods to determine it.[10] Moreover, ISO isn't 'sensitivity', although it is tempting to think of it that way. Sensitivity depends on the size and performance of the photodiode, including its efficiency: that is, how effectively it accumulates charge in response to light. Changing ISO only alters what occurs downstream: how much the signal is amplified before it is converted to a digital value, and additional post-processing used for lightening the image.

Such processing comes at a cost, such as loss of dynamic range. Every sensor has a maximum output or 'well capacity': the number of electrons it can accumulate (say, 40,000); and a minimum output: the number of electrons at its *noise floor* (say, 10). The ratio of these is the dynamic range; in our example, it is 4,000. In order to digitize the output, the voltage (proportional to the number of electrons collected) is fed into an ADC (analog-to-digital converter) with some *bit depth*. We encountered this with digital audio, and here, a bit depth of 12 would be appropriate because it allows $2^{12} = 4,096$ values, comparable to the native dynamic range of the sensor. This is an optimal state of affairs because the digital output is effectively counting single electrons. But if the sensor is producing a low voltage (due to a paucity of light, for example) and we use ISO to amplify it by, say 10× before it is digitized, then every step in our output corresponds not to one, but to 10 electrons. This means 10× lower resolution, introducing a kind of 'graininess' in the gray-scale resolution.

Cameras are often advertised in terms of their megapixels: the total number of sensors that comprise the array. There is a tacit assumption that more pixels are better, but this isn't always the case. Certainly, if the count is low because the photodiodes are extremely large, this will limit the spatial resolution, leading to pixelated images. But in making them *too small*, we are back to the issue that ISO must aim to correct for. Ideally a pixel should be small enough for sufficient resolution, but large enough to give high dynamic range. This is a moving target, because newer and more efficient sensors can be made smaller without much performance loss.[11]

A full-frame sensor, which mimics the size of a 35 mm film exposure, has a total area of about 860 mm^2. If it holds 24 megapixels, each pixel will measure about 6 microns on a side. The same megapixel count applied to a cell phone camera sensor—which might have a total area of 25 mm^2—will have sensor dimensions of about 1 micron, or about 36 times less light collecting area and far less signal to work with. If we reach a high megapixel count due to a larger overall sensor, we gain the advantage that it can capture a wider field of view, but at the cost of a larger camera size, of course. As a general rule, better performance means a bulkier device.

The evolution of the digital camera has been rapid, and this digression has only touched on some of the fundamentals. We will investigate lenses more in the following chapter, in the context of binoculars and telescopes. Here, we will finish with a look at the most extreme optical instruments that nature has engineered: the eyes of birds.

Avian vision and its capabilities

The prototypical camera we've become familiar with is similar in many ways to an eye. Both exploit refraction and redistribute light across a network of sensors in order to form an image. So, like cameras, eyes must negotiate various performance trade-offs. This leads to a gamut of designs, even just within the class *Aves*, in order to deal with the variety of environments and strategies for foraging and survival. Here, we'll survey the physics behind some famous (and not-so-famous) aspects of avian vision.

Raptors are celebrated for their visual acuity, and we know that the eyesight of various hawks and eagles exceeds our own in several ways. Yet, knowledge

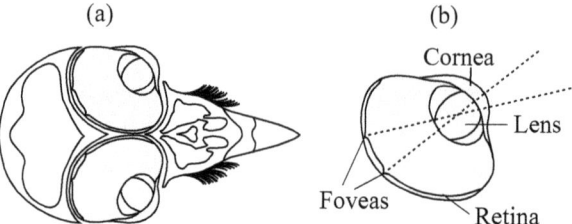

Figure 9.13 (a) Cross-section of the skull of a Broad-winged Hawk *Buteo platypterus* from above, showing the full extent of the eye volume (shaded). (b) Several key components of the eye are highlighted: the cornea and lens, which admit and focus light; and the retina, which includes two foveas, which appear as indentations in the retina. The dashed lines indicate directions of incoming light that falls upon the foveas. [Modified from Wood (1917).[12]]

here is limited; a literature review in 2020 found that vision has been described for less than 2% of all raptor species.[13] Extreme visual capabilities are certainly not universal among them, with some species having less acuity than humans.[14] Exceptional vision comes at a high biological price, and only those that hunt from on high need pay it. Terrestrial, opportunistic raptor species can get by with less.

The most obvious way to improve an eye is to make it as large as possible. Our eyes take up about 5% of cranial volume, but in some raptors they account for 50% of the space.[15] Eagle or hawk eyes may not appear especially large, but we are only seeing a portion of them, as is evident from diagram (a) of Figure 9.13.

As with larger cameras, larger eyes have more sensor area, a wider aperture to admit more light, and longer focal lengths, which make for larger images and greater resolution. Human eyes have a nominal focal length of about 25 mm; for some raptors, it exceeds 30 mm.[16] Another strategy to improve acuity is to pack more light-sensitive cells into certain parts of the retina. These regions, the *foveas*, are shown in diagram (b) of Figure 9.13, where they appear as small indentations in the retina. The dashed lines indicate directions of light, passing through the cornea and lens, that fall upon these regions.

Most birds, like humans, have only a single fovea, corresponding to the center of the visual field. The second fovea is an adaptation used by certain prey-pursuing raptors (and even a few non-raptors, such as some terns).[17] It provides improved binocular vision in a forward direction, simultaneous with the more acute side-vision afforded by the central fovea.[18]

Typically, the fovea is a few square millimeters in extent and, unlike the rest of the retina, contains few low-light sensitive rod cells. Instead, it holds a high density of cones—about 160,000 cells per square millimeter for humans, and over 450,000 for some raptors, endowing the latter with well over double the resolution.[19] Both bird and human fovea employ only two, instead of the

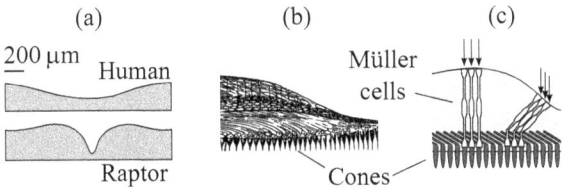

Figure 9.14 (a) Relative size and shape of human and Black Kite *Milvus migrans* fovea cross-sections. (b) A sketch of the cross-section of the human fovea. Cone cells lie along the bottom and are covered by layers, which include vascular tissue, and nerve and ganglion cells connecting to the cones. (c) A schematic of an avian fovea highlighting the presence of Müller cells. It has been proposed that these cells act as optical fibers directing light to the cones beneath. In that case, the steep walls of the fovea would enable greater resolution without having to make the cone cells smaller. The arrows indicate incoming light. [Diagram (a) adapted from Frey, Zimmerling, Scheibe, *et al.* (2017)[20] and Potier, Mitkus, Lisney *et al.* (2020).[21] Diagram (b) reproduced from Quain, Schafer, Sharpey *et al.* (1878).[22] Diagram (c) modified from Zueva, Makarov, Zayas-Santiago *et al.* (2014).[23]]

normal three (or four) types of cones here, trading some color discrimination for better resolution.[24]

The pit-like shape of the fovea, which is apparent by taking a cross-section of the retina, is notably different in structure for raptors and humans, as shown in diagram (a) of Figure 9.14. In our case, the gentle depression results from a localized reduction in the blood vessels that lie atop most of the retina.[25] These comprise the upper layer of tissue of diagram (b), which becomes markedly reduced to form the 'pit'. Thinning or eliminating material here aids vision by reducing light losses from absorption or scattering.[26] Also thinned out is the layer containing the *ganglion* cells, which lies atop the cones and provides the connections between cells and the optic nerve.

Diagram (c) shows a schematic cross-section of a raptor fovea, which again features cone cells beneath nerve tissue and a localized depression in which the layers above the cones are thinned. Highlighted are specialized *Müller cells*, whose function we will consider momentarily.

Various explanations have been offered for the unique shape of the raptor fovea with its steep walls.[27] It certainly has the effect of reducing the amount of tissue that light must traverse, at least for the central cells. Some researchers have postulated that the shape helps magnify the image, but the expected effect is small. Others suggest that it creates a kind of distortion that allows movement to be more readily noticed, but nothing definitive has been demonstrated.[28] The Müller cells shown in Figure 9.14 (c) may exploit the geometry by acting something like fiber optics that redirect incident light.[29] As shown in the figure, the bundle of cells that lie on the sides of the fovea present a narrower cross-section to the oncoming light, which would allow for greater resolution than the native cones could have, or if the Müller cells terminated on a flat surface, as do those on the left, outside the fovea.

We can estimate acuity by scrutinizing the cone size and density, but there are behavioral ways to quantify it as well. An image with alternating black and white stripes is distinguishable from a solid gray pattern, up to the point where the widths of those stripes are made so small that they can no longer be resolved. Acuity can be ascertained as the maximum number of 'cycles' (line pairs) per degree that do not appear gray. Fortunately, birds can be trained to respond differently to striped and solid images, and this provides a way for us to assess their vision.[30] Using such methods, acuities as high as 143 cycles/degree have been measured for eagles, which is over twice our normal acuity of 60 cycles/degree ('20/20' vision).[31] This is consistent with the difference in cone densities.

Predatory diurnal raptors also improve acuity by minimizing diffraction via keeping their pupils open wide, even under bright conditions. (Diffraction was introduced in the context of audio, when we considered how sound waves spread out when they pass through a relatively narrow opening. The same occurs with light, which is another drawback to passing it through a narrower aperture.) The excess glare is sometimes reduced by the presence of dark malar feathers; these marks function like the black smudges that some athletes apply under their eyes. The prominent brows that give various raptors their fierce appearance also help in this regard by blocking sunlight. The resulting overhead blind spot helps explain the paradox of how such keen-sighted birds are yet so vulnerable to collisions with cables and windmills.

With only 20–25% of their retinal cells being rods, diurnal raptors rapidly lose vision as light levels drop. For owls, unsurprisingly, rods account for some 90% of the receptors.[32] With their acute hearing, they can afford to trade visual resolution for improved low-light capabilities. Their tubular eye shape helps them employ large, light-gathering apertures.

Other nocturnal birds, such as nightjars and potoos, are famously recognizable due to their eye-shine, revealing another widely used adaptation for seeing in the dark. Because some amount of incident light is not absorbed by photopigments, a reflective layer behind the retina directs it back through so that it has a second opportunity to be sensed, thereby improving sensory efficiency.[33]

An oft-overlooked example of birds that capture elusive prey in very dark conditions are those that forage deep under water, beyond the reach of most sunlight. Consider that 200 meters down, even in clear water, the light level diminishes by a factor of 100. Add to that the shallow solar incidence intrinsic to a polar region, and it is astonishing to learn that Emperor Penguins *Aptenodytes forsteri* have been found to forage at depths of more than 560 meters, and at

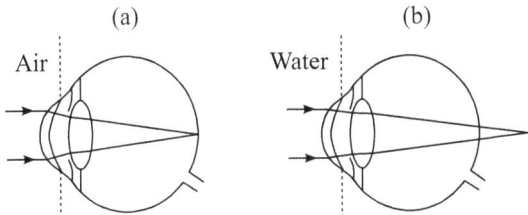

Figure 9.15 (a) Schematic of a human eye viewing a distant object through the air. The refraction at the cornea and lens produces a focused image on the retina. (b) When the eye is in water, less refraction occurs at the cornea because the change in refractive index is less. The focus is now beyond the retina, as the lens cannot be altered in shape sufficiently to accommodate the change.

over 440-meter depths during the sunless winter. Such conditions, at best, would amount to starlight illumination on a moonless night.[34]

Low-light conditions or not, diving birds have the additional challenge of seeing well while immersed. We humans learn quickly, upon learning to swim, that our unaided eyes—designed for life out of water—yield hopelessly blurry images when submerged. How do birds manage to see well both out of and under water?

First, we must understand why our eyes function poorly when submerged. In describing the basic eye/camera device in this chapter, we've simply referred to a single lens as the mechanism for redirecting light to form an image. For the eye, the lens is a curved element (set behind the pupil) whose shape can be altered by muscles, but it isn't where the bulk of the refraction occurs. That takes place in the cornea, which is more than a protective cover. It does much to reshape the light, while the lens provides fine-tuning by changing focal length to accommodate objects at different distances.

In Figure 9.15 (a), we illustrate light paths into an eye in an air environment. Recall that, to bend light, it isn't the refractive index of a material that matters so much as the *change in index* at an interface. Our corneas have an index of about 1.3 to 1.4; this is sufficiently high that in air (where $n = 1$), its curved shape is well suited to begin the focusing process.

As shown in (b), however, this effect is effectively eliminated if we replace the air with water, which has an index of 1.33. The loss in refractive power produces a focal length too long for our eyes; and even as the lens attempts to compensate, it cannot, leaving the scene out of focus. Donning goggles immediately solves this problem by placing a region of air between our corneas and the water, giving them back their refractive capability.

For an example of how some birds solve this problem, we can look to Hooded Mergansers *Lophodytes cucullatus*, which have been studied extensively

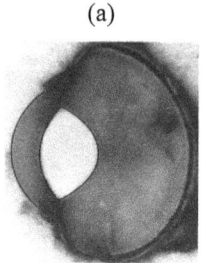 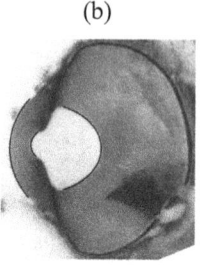

Figure 9.16 (a) Cross-section of the eye of a Hooded Merganser *Lophodytes cucullatus* in its normal state, in air. (b) Cross-section of the same eye, showing the change in the lens which occurs when the bird is submerged in water. The bulging shape of the lens is apparent, and will compensate for the loss in refraction due to the water–cornea boundary. [Reproduced from Urban, Uwurukundo, Stumpf *et al.* (2020) under the terms of the Creative Commons CC-BY license.[35]]

in terms of their visual adaptations for foraging underwater.[36] It is estimated that their corneas account for some two thirds of the refraction when out of water, but less than 10% when submerged. To make up for this, the lens undergoes a significant shape transformation in which it is pushed up against the iris so as to bulge through the pupil. It gains significantly more curvature, as seen by comparing Figure 9.16 diagrams (a) and (b), which show the merganser eye in air and water, respectively.

This appears to produce refractive power that effectively compensates for the loss of corneal refraction. Note that 'power' here does not mean magnification, but the ability to *accommodate* or focus. This is expressed in *diopters*, equal to $1/f$, with the focal length f measured in meters. Mergansers can accommodate by an astonishing *90 diopters* in just one second; human accommodation might be around 15 diopters, for children. We steadily lose this capability, typically having only a range of about 1 diopter by the time we are in our 50s.[37]

Our study of cameras and eyes began simply: with an array of sensors behind a single lens. This has sufficed to explain their basic operation, but as we're seeing now, a series of refractive interfaces (such as cornea and lens) making a compound optical system can have greater utility. Any non-trivial device for further manipulating images will require multiple optical components, and in the following chapter we will see how to arrange them to build the indispensable tools of birding: telescopes, camera lenses and binoculars.

Chapter 10

From a Great Distance: Lenses, Binoculars and Scopes

Birding may take place over large distances, including the occasionally vast space between us and our quarry. We should be thankful to live in an era in which affordable optical instruments make viewing far-off birds possible. It is a gift unimaginable to most of the humanity that has gone before us. In this chapter, we'll explore the basic operation of these devices that we are liable to take for granted, even as they make the distant magically appear so close.

Lenses and images

In the previous chapter, we showed how a transparent material with appropriately curved surfaces—a *lens*—provides an elegant solution to the problem of admitting more light into an eye (or camera) while maintaining focus. Humans have long used this simple tool for other purposes; with its ability to intensify sunlight, starting fires may have been its earliest practical application. The Greek writer Aristophanes referred to a 'burning-glass' in 424 BCE, and lenses have been recovered from ruins dating back hundreds of years prior.[1] Moreover, placing such a device between the eye and a nearby object makes its ability to *magnify* readily apparent. This enables fine-scale inspection and manipulation beyond the native range of the unaided eye, and can correct for farsighted vision (as discussed below).

Whereas a single lens is useful, combinations of them can produce instruments capable of seemingly magical results. As birders, we usually take for granted that a few pieces of glass, properly arranged, can bring the birds right next to us and make details pop out in spectacular fashion. Here, we will work through, lens by lens, how this alchemy of light and images is performed. We've already considered a number of lens examples in the

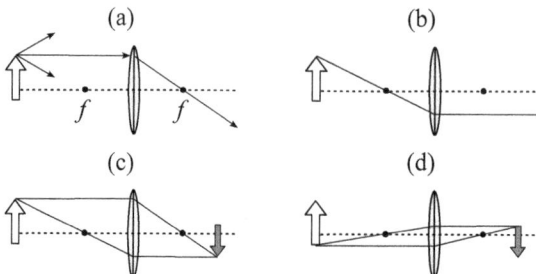

Figure 10.1 (a) Light from a point on an object radiates outward, as indicated by rays. The ray that meets a positive lens in a perpendicular fashion will, by definition, be directed to pass through the focal point beyond; a distance *f* from the lens. (b) A ray from the object passing through the closer focal point will leave the lens in a perpendicular direction. (c) The intersection of the two rays from (a) and (b) indicates where the image of the top of the object appears. (d) A similar construction is used to show the image from the bottom of the object.

previous chapter; now we'll need to refine a few ideas and build upon that knowledge.

Let's start with a typical optics diagram involving an object to the left of a single lens, as shown in Figure 10.1 (a). The lens has focal points marked *f*, which, we recall, are locations where parallel rays entering the other side would converge. We illustrate a variety of rays emanating from the top point of the object, and follow one ray, which is moving parallel to the optical (horizontal) axis. Rays arriving along this direction must go through the far focal point, as shown. In (b) we select a second ray, which specifically passes through the focal point on the left side of the lens. Upon reaching the lens, it must become parallel to the axis. In (c) we show both of these rays and their intersection, which marks the location where the light from this point is focused. In (d) we show the results for similar constructions of other points.

If the object in Figure 10.1 is slowly brought closer to the lens, the image will move further away. We saw this when we constructed Figure 9.12, in the context of *depth of field*. As the object nears the focal point *f*, the image will eventually be infinitely far to the right. But what happens if the object is brought even closer than *f*? In Figure 10.2, we have made the ray diagrams that will let us answer this.

The steps taken in the ray-tracing diagram (a) are identical as before; what is different is that the two rays from the tip of the object can only intersect *on the same side as the object*. This does result in an image, but it can be recognized by looking back through the lens, as shown in diagram (b). This is said to be a *virtual* image, as opposed to a *real* one, but it is no less authentic an effect. A real image is one that can be projected onto a surface placed at that location.

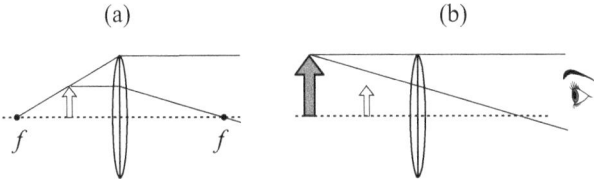

Figure 10.2 (a) An object is brought close to the lens, such that the distance is less than a focal length indicated by f. A ray passing through the focal point and lens, and which also passes through a point on the top of the object, is shown, as well as a ray that meets the lens perpendicularly and which must pass through the other focal point f. (b) An observer on the other side of the lens will observe these rays as if they are coming from another point in space, left of the lens, corresponding to the virtual image of that point. In this case, the image for the object will appear larger than the object itself, demonstrating the basic operation of a magnifying glass.

That surface may be a retina, a piece of film, a digital sensor or a distant movie screen, for example. A virtual image itself cannot make such a display; it may not even be physically possible to place a screen where the image is located. One's reflection in a flat mirror is a virtual image, standing an equal distance *behind* the surface, where it is utterly unreachable.

As Figure 10.2 makes clear, the image will appear larger than the object, so the lens acts as a simple magnifying glass. We can quantify this by taking the ratio of the angular extent of the image to that of the object. Besides enlarging nearby objects, this optical system can also correct for farsightedness, as shown in Figure 10.3. Diagram (a) shows how a *hyperopic* eye regards a distant object; the focal length is appropriate for making a sharp image on the retina. But for a closer object, as in (b), the focus moves further back, and the lens cannot adjust enough to account for it. The light reaching the retina is out of focus. In (c), a convex lens stands near the eye, providing the additional converging power and enabling an image to be formed on the retina.

For the nearsighted (myopic) eye, the problem is that the lens in its relaxed state has too short a focal length. Bringing an object up close pushes the image

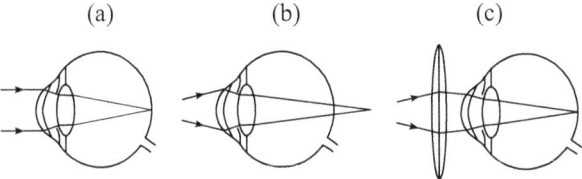

Figure 10.3 (a) A farsighted eye is able to focus the light from a distant object on the retina. (b) For a closer object, the arriving rays are divergent, which causes the focal point of the image to move backward, and the eye is unable to modify the shape of its lens (shaded) to accommodate for it. (c) Placing a positive lens between the object and the eye brings the rays into a more parallel configuration, allowing for the focus to be brought back to the retina.

152 • The Physics of Birds and Birding

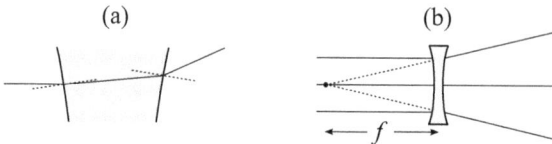

Figure 10.4 (a) Close-up of a negative lens, showing how a ray (solid line) meets the two surfaces relative to the normal direction (dotted lines). Knowledge of refraction via Snell's law immediately shows how the ray will be redirected. (b) Three parallel rays passing through such a lens diverge.

further back and onto the retina, but distant objects will focus at a point in the middle of the eye. What is needed is a device to do the opposite of what our convex lens does: it must cause the light to diverge rather than converge.

This is accomplished with a *concave* lens, as shown in Figure 10.4. In (a), we trace a ray coming in near the top of the lens. Applying Snell's law at each surface shows how it will deviate away from the incoming direction. A full set of parallel rays is shown in (b). Although this lens makes the rays diverge rather than converge, the idea of a focal point still applies; it simply lies on the same side as the incoming rays. To an observer on the right, the light appears to be coming not from a distant source, but from a point a distance f away.

Clearly, placing such a diverging (or negative) lens near the eye will have the reverse effect from that of the converging (or positive) lens in Figure 10.4; and with the correct focal length, distant objects will be brought into focus. While it cannot function as a burning-glass, a negative lens nonetheless has great utility, and it was known of as early as the thirteenth century, the same era in which eyeglasses were first described.

Over the next several hundred years, in which primitive spectacles were becoming more common, it would be inevitable that a glimpse through several lenses in a fortuitous arrangement might occur, and such an arrangement allows for *distant* objects to be magnified. The utility of such a device went widely unrecognized until 1608, when the Dutch eyeglass maker Hans Lippershey filed for a patent for his 'looker', the name he gave his telescope.[2] Suddenly, the invention was in great demand, primarily because of the obvious advantages it could confer in military situations. Galileo would grind his own lenses and turn his first modest instrument, with a magnification of just 3x, to the skies in 1609. Few actions in the history of science were more far-reaching.

Tracing through a ray diagram for two lenses will allow us to understand how a telescope achieves its magnification. A distant object (that isn't a point source like a star) has some spatial extent that can be characterized by an angle α, as shown in diagram (a) of Figure 10.5. Each end of the object produces rays that arrive from slightly different directions. In (b), we consider the trajectory

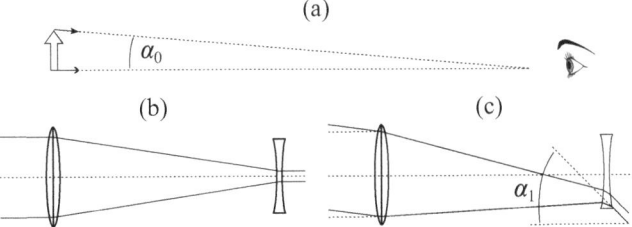

Figure 10.5 (a) A distant object has an angular extent of α_0. (b) The effect on rays from the bottom of the object, which arrive perpendicular to a pair of lenses. The left lens—the objective—is positive and makes the rays converge, while the rightmost lens—the eyepiece—makes the rays parallel. (c) Light from off-axis, at an angle α_0, enters the same system of lenses. Passing through the eyepiece, it undergoes a large angular deviation, α_1 relative to the horizontal. The image therefore takes up a larger angular extent than that of the object seen with the unaided eye.

of the light from the base of the object, which lies along the optical axis for a pair of lenses, the first convex (positive) and the second concave (negative). The second lens is placed closer than the focal length, so the converging rays from the first lens are now made to diverge by an equal and opposite amount, resulting in parallel rays to the right of this lens. An observer here will see the light as coming from the base of the object.

In (c), we trace the light coming from the top of the object, which strikes the first lenses at an angle α_0. This light is bent downward, and if not for the second lens, it would produce the tip of an inverted real image. But this off-axis light is intercepted and bent to an even larger degree. For the observer, the light originating from the top of the object now appears to be coming in at a much larger angle, α_1. This new larger angle, divided by the original, gives the apparent magnification, $M = \alpha_1/\alpha_0$.

This is a very basic design and it has many drawbacks, but one can still purchase opera glasses that use this simple approach. Most telescopes and binoculars have become far more complex, and we will see how they have changed and why, as we look into more details of their performance.

Galileo's and Kepler's telescopes

After constructing a few lens diagrams, a useful result for tracing rays becomes apparent. As shown in Figure 10.6, drawing rays parallel to the axis and passing through the focal points creates a trapezoid. But a line from the object passing through the center of the lens must also arrive at the same image point.

With this in mind, we'll sketch a diagram in Figure 10.7 for our telescope to estimate how much magnification it provides. Diagram (a) shows only the

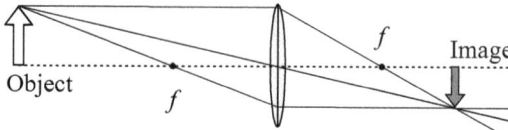

Figure 10.6 A typical ray-tracing diagram for a positive lens. Rays from a point on the object diverge and meet the lens at various angles. The top ray arrives perpendicular to the lens and thus must pass through the focal point. The ray passing through the center of the lens is undeviated. The ray passing through the near focal point will produce a ray perpendicular to the lens after it exits. The intersection of these rays marks the location of the image.

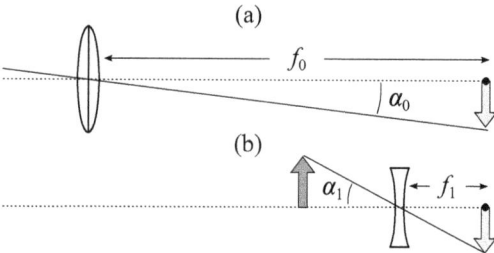

Figure 10.7 (a) Ray diagram for the Galilean telescope objective lens alone with an object at infinity. (b) The eyepiece is placed such that it is one focal length (f_1) from the focal point of the objective (f_0). The resulting virtual image will appear to lie between the lenses with an angular extent of a_1.

convex *objective lens* and how it alone would form the image of a distant object at f_0, with height h and angular extent a_0. In (b) we place the concave eyepiece, or '*ocular*', such that its focal point, f_1 coincides with that of the objective.

If these angles are relatively small, we can approximate them as $a_0 = h/f_0$ and $a_1 = h/f_1$. With magnification defined as the ratio of angular sizes, $M = a_0/a_1$, we have an elegant result:

$$M = f_0/f_1$$

Light from a distant point enters the telescope in the form of a cylinder of light, or beam, which is as wide as the objective, and then exits the eyepiece as a much narrower beam of parallel rays; refer to diagram (b) of Figure 10.5. The ratio of these beam widths is also M. The telescope is essentially a funnel that accepts a wide cross-sectional area of light and compresses its width. Ideally, the exiting beam will match the pupil diameter of the eye behind the scope.

We use the general term *aperture stop* to describe whatever determines the width of the light beam entering the device; for our telescope, it is the rim of the objective lens. In a typical camera, it is the iris diaphragm, located behind the outermost lens. Looking through the instrument, the aperture stop defines

From a Great Distance: Lenses, Binoculars and Scopes • 155

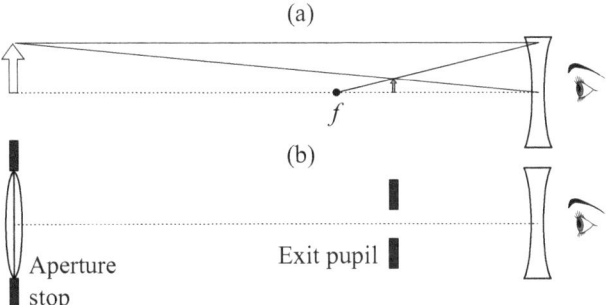

Figure 10.8 (a) Virtual image (gray arrow) construction for a nearby object (white arrow) beyond the focal length *f* of a negative lens, based on two intersecting rays. (b) Application of the same result when the object is the aperture stop encircling the objective lens of a Galilean telescope, which produces an image defined as the exit pupil.

the limit: that is, the angular extent that can be viewed. The image of the aperture stop, as seen through the eyepiece, is called the *exit pupil*.

Figure 10.8 shows how we can find the exit pupil for our scope. In (a) we construct the virtual image for the eyepiece alone, for an object located where the objective would be. Now imagine the object isn't an arrow, but a circular ring defining the outer rim of the objective; as shown in (b). The symmetry of the system lets us immediately draw the full exit pupil, with the edge of the objective lens appearing as a circle inside the scope.

This is not an optimal configuration for the exit pupil, as it contributes to a narrow *field of view* (FOV) among other things. FOV is the angular extent of the scene that can be taken in, and typically is not as simple to determine as magnification. Here, we will make some approximations to gain some intuition for the factors affecting FOV.

For our telescope, FOV is determined by the requirement that light pass through both the aperture stop (the objective lens) and another constriction—referred to as the *field stop*—which here is the entrance pupil of the eye. The best (but uncomfortable!) case would be to get the eye as close to the exit pupil as possible, right up against the eyepiece. Figure 10.9 shows this geometry for a ray entering the objective at an angle a_0, such that half the cylinder of light entering the scope cannot enter the eye; this is indicated by the shaded region. This loss of light that comes with increasing the angle is *vignetting*, which causes a familiar, circular fading pattern.

In diagram (b), we estimate the total FOV in terms of the angle F. Because the situation in (a) depicts the top limit of the FOV, F must be twice the size of a_0, which is the deviation of the central ray from diagram (a). How do we find this angle? A full derivation is beyond the scope of this book, but

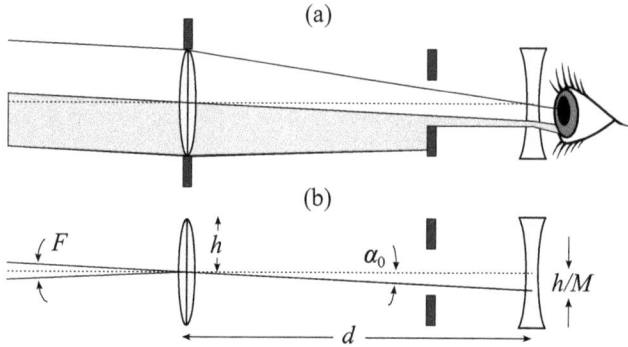

Figure 10.9 (a) Off-axis light enters a Galilean telescope at such an angle that half the light, indicated by the shaded region, cannot enter the observer's pupil. (b) A geometric construction of the same scenario, showing only the ray passing straight through the lens center at an angle α_0. The objective lens has a radius h and is a distance d from the eyepiece, and the deviation of the ray at the eyepiece is h/M. Because a similar ray from below would admit the same amount of light, the angle $F = 2\alpha_0$ corresponds to the full field of view (FOV).

we can make some approximations to simplify matters. Since magnification captures the ratio of the beams' widths entering and leaving, downward deviation should be roughly the objective radius h divided by M. With a distance between the lenses d, we can say that the angle is defined by $\tan(\alpha_0)$ = h/Md, where 'tan' stands for tangent: the ratio of the opposite and adjacent sides of the triangle made by α_0. If the angle is small, we can approximate this as $F/2 = \alpha_0 = h/Md$. Thus, this scope quickly loses FOV if we increase its magnification, unless we use a larger objective or bring the lenses closer by reducing focal lengths.

Also note from diagram (a) that, if we move the eye away from the eyepiece, it will see even less of the light on the periphery; that is, the FOV will fall off rapidly as the eye pulls back. This design therefore has very poor (zero, in fact) *eye relief*, defined as the maximum distance from the eyepiece that provides the full FOV.

If the telescope's exit pupil lies beyond the eyepiece, where it may coincide with the entrance to the eye, the eye relief will be greatly improved. Fortunately, there is a design tweak that provides just that, which was discovered by the great astronomer Johannes Kepler in 1611. He replaced the negative eyepiece with a positive one, positioned beyond the focal point of the objective lens, as shown in Figure 10.10.

Tracing through diagram (a) for rays coming from just above horizontal, we note how the rays cross and that, to the right of the eyepiece, the light will appear to come from below horizontal. This means the image will now be upside down. In diagram (b) we constructed the exit pupil using the same

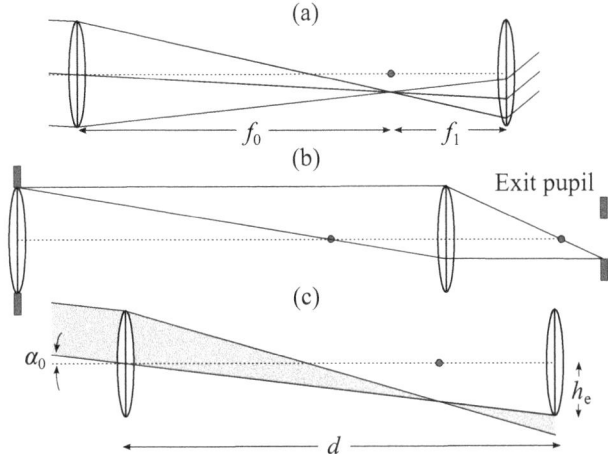

Figure 10.10 (a) The basic Keplerian telescope design uses two positive lenses, with the eyepiece placed one focal length f_1 behind the focal length of the objective, f_0. Parallel rays arriving from just above-axis result in parallel rays leaving the eyepiece with a large deviation (indicating magnification) but are inverted, meaning that the image will be upside down. (b) The exit pupil is found by constructing the image of the aperture as made by the eyepiece alone. Unlike the Galilean case, the exit pupil here lies to the right of the eyepiece. (c) The construction for finding the FOV involves an incident angle α_0 for which half the light (shaded region) will fall outside the radius h_e of the eyepiece.

method as before: by forming the image of the aperture stop made by the eyepiece; and this image lies outside the telescope. In diagram (c) we trace the path of rays corresponding to 50% vignetting in order to estimate the FOV as we did before. Using a similar approach as in Figure 10.9 (b), the angular deviation can be expressed as $\tan(F/2) = h_e/d$ where h_e is the radius of the eyepiece, which defines the field stop for this case.

Let's run some numbers for two similar telescopes of Galilean and Keplerian design, both with 25 mm lenses, an objective focal length corresponding to $f/8$ and 10× magnification. Overall, they would have comparable lengths of 180 and 220 mm, respectively; but their FOVs would differ dramatically, roughly 1.6° vs. 13°. Moreover, the difference in eye relief makes the Keplerian scope far more pleasant to use.

But there is always a price for performance gains, and here it is the upside-down image. For astronomical use, this makes no difference, since celestial objects have no 'right side up'. For terrestrial use, it is very off-putting, not only because the world is turned on its head, but because moving the scope causes the image to run off in the opposite direction to what was expected. We can mitigate the inversion by inserting a third lens between the objective and ocular, which provides another reversal. The trade-off here is that the overall length must increase, leading to the classic spyglass we might imagine an

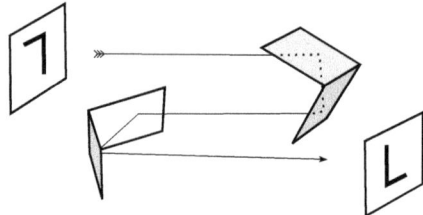

Figure 10.11 An upside-down and left–right reversed object (an 'L') is on the left. Light from this object is reflected four times, with one pair of reflections inverting left and right and the other flipping the up and down directions, resulting in a complete image reversal.

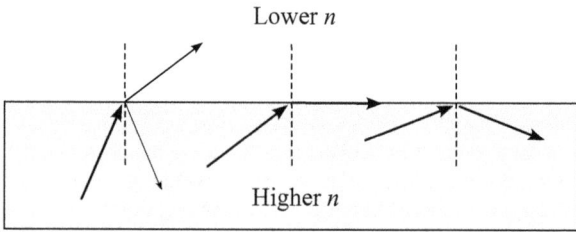

Figure 10.12 Light within a high-index material (shaded region) approaches a lower-index region above it. On the left, the incidence angle is smallest, and most light passes through and gains a large refraction angle, while some light is reflected. In the center diagram, the light arrives at the critical angle, in which the refracted light leaves parallel to the surface. On the right, we have exceeded this critical angle, for which transmission is no longer possible, and total internal reflection (TIR) must result.

ancient mariner wielding. If we want something compact and more amenable for use in binoculars, we'll need to go a different route.

Reflections, of course, flip images in addition to redirecting light. With the clever arrangement of four reflecting surfaces, we can invert the up–down and left–right sense of a beam without changing its direction, although it may shift the optical axis. Figure 10.11 illustrates how to do this.

These reflections are best made by using the inner surfaces of glass prisms, which leverage the *total internal reflection* (TIR) that occurs at the boundary of materials with different refractive indices. The origin of TIR can be seen by considering the path that light takes when meeting a transition from higher to lower index. Figure 10.12 shows paths for several rays at different angles. Applying Snell's law, we find that as the incident angle increases, the transmission angle grows rapidly until we reach a critical condition where the reflected light moves parallel to the surface. Beyond this angle, the light cannot pass through, and instead must be wholly reflected back inside the high-index medium.

The critical angle at which the reflection becomes total will depend on the difference in refractive indices. The larger the difference, the less steep an angle is required. With the proper materials, then, we can construct a system that performs the four reflections shown in Figure 10.11 using a single unit that cannot go out of alignment, as independent mirrors could. This device is the *Porro prism*, developed in the mid-nineteenth century and still widely used. Its presence in a pair of binoculars is belied by a 'bent' shape to the housing, such that eyepiece and objective do not lie upon the same axis. So, by folding the light, we harvest the gains of a Keplerian design and deliver a non-inverted image and compact shape amenable to hand-held use.

Having constructed a prototype of arguably the most important tool of birding, we can now explore enhancements, pitfalls and their solutions in modern optical instruments.

Optics and their limitations

We now have a blueprint for devices that produce magnified, non-inverted images with relatively wide fields of view; while being easy to wield and comfortable for our eyes. Yet, we are a long way from modern telescopes and binoculars. Some details we have yet to cover are obvious; others are subtle, concealed within the instrument.

A conspicuous difference among binoculars is the form of the optical axis; they have either the familiar Z-shape due to the Porro prisms, or they manage to keep the light confined within a straight tube. The latter makes for more compact optics, but require sophisticated and, usually, more expensive components.

The straight-through design relies on a *roof prism*; a clever, precision glass sculpture that performs the same work as a Porro, but within a much tighter space. There are various types of roof prisms, but they all share a similar feature: a pair of reflecting surfaces at right angles, resembling the roof of a house.

To appreciate the artfulness of this shape, consider first the reflection from a flat mirror, as shown in Figure 10.13 (a). Tracing the rays, which reflect at the same angle at which they strike, shows how spatial orientation is maintained. In (b) we use two mirrors at 90°, and again trace the light. The additional reflection causes the objects to appear flipped. In (c), we show the simple *Amici prism*, where two surfaces form a roof on the long (left) side of a triangular block.

The Amici prism alone will not do everything we need. If it provides one flip, say from left to right, it won't perform the up–down inversion. We will need more reflections; but again, we wish to arrange matters so that the beam has no net deviation, as the Porro causes. Several clever means to do this have been invented, such as the *Abbe-König prism*, shown in Figure 10.14 (a). Here,

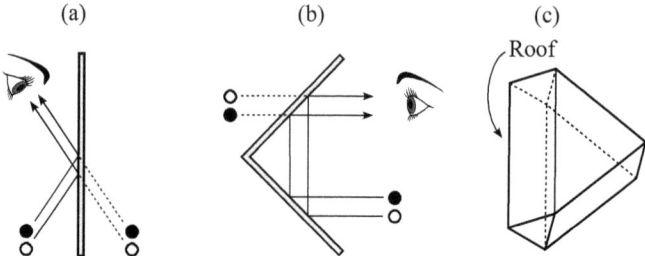

Figure 10.13 (a) A single reflection does not invert an image, as the orientation of the two points remains the same. (b) A double reflection made from two mirrors at right angles causes the image to be inverted. (c) A simple Amici roof prism is essentially a triangular block of glass on which one side is replaced by a pair of surfaces at right angles.

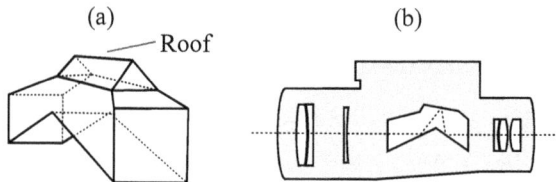

Figure 10.14 (a) The Abbe-König prism features a roof placed atop two total internal reflection surfaces, leading to four reflections for light passing through it. (b) Incorporation of the prism in a binocular body between various lens elements.

the long roof section sits above two inclined planes, each oriented at 30°. These act to flip the image vertically, while the roof takes care of the left–right reversal. And, unlike a Porro, the beam emerges along the incident axis.

As might be expected, the performance of this prism will be compromised if the surface geometries do not meet strict tolerance controls. Manufacturing such precision devices is not trivial. Typically, two pieces are fashioned and then joined with optically clear cement. Their incorporation into binoculars is shown in (b).

The roof strategy is brilliant, but it is victim to an arcane, performance-degrading effect that Porros are immune to. The division of light into two halves that reflect off different surfaces results in destructive interference that compromises the image quality. Specialized treatment of the roof surfaces, known as *phase coating* or *p-coating*, can effectively mitigate this complication. The physics is not simple.[3]

Another roof prism employed in many binoculars is the *Schmidt–Pechan*, shown in Figure 10.15. The side view (a) shows the light path; the lower prism reflects the light twice, directing it to the roof prism above. Note that the second reflection occurs at a fairly steep angle such that TIR is not achieved, so some amount of light escapes out of this surface if a reflective coating isn't applied here.

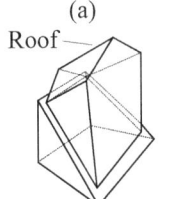
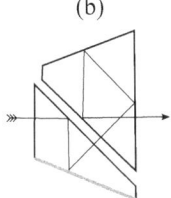

Figure 10.15 (a) The more compact Schmidt–Pechan prism features two wedge-like prisms, the top having a roof. (b) A ray diagram showing that a total of six reflections (two on the roof) occur with this device. The reflection at the bottom surface (gray) is not steep enough for TIR, and so a reflective coating is required here.

Metals come to mind immediately for reflective coatings. Their intrinsic shininess stems from their untethered electrons, easily set into motion by incident light and free to re-radiate. This isn't entirely efficient, though, as some energy is absorbed. Typical materials reflect from 85% to 98% (aluminum and silver, respectively) of the incident light. The latter may seem adequate, but those last few percentage points make a difference in low light conditions.

The best way to enhance reflectivity is by using the same mechanism as iridescent feathers, which shine brilliantly without metal. A hummingbird gorget uses repeated layers with differing refractive indices and exploits constructive interference. Similarly, a series of thin dielectric (that is, non-metallic) coatings with appropriate thicknesses can be engineered to have reflectivities greater than 99% over the entire visible range.[4]

Aside from certain prism surfaces, reflection is a liability for optical components, and considerable effort goes into reducing it. When incident light strikes a lens, some (typically 4% for glass) bounces off due to the change in index. Again, this may not seem like much, but since multiple surfaces exist in any instrument, there is a cumulative effect from these losses. The best optics take measures to mitigate reflections.

The fraction of light power reflected from an interface defined by different refractive indices n_1 and n_2 is captured by the *reflectance* R. This is a simple function of their difference and sum, namely:[5]

$$R = \frac{(n_2 - n_1)^2}{(n_2 + n_1)^2}$$

For air and glass (n_1=1, n_2=1.5), this works out to 0.04. If another material, with an intermediate index, is placed between the air and glass, we will get two new reflections; this seems like a bad idea, until we calculate their reflection coefficients. Choosing an index of 1.25, for example, the first reflection coefficient

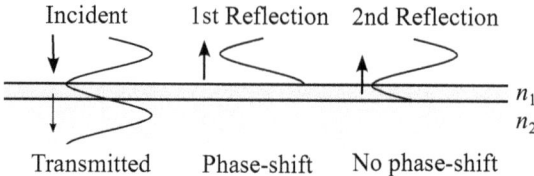

Figure 10.16 Incident light from above reaches a structure with two different refractive indices, $n_1 > n_2$. The left diagram shows that the incident and transmitted wave is continuous across the boundaries. The middle diagram shows the reflection from the top surface, which is inverted relative to the incoming wave, because the reflection is from a boundary into a slower medium. On the right is the reflection from the middle surface, which is not inverted, because the transition is from a slower to a faster medium.

is 0.012 and second is about 0.008. This simple treatment, therefore, reduces the total reflected light by half.

We can do even better by choosing the new layer thickness to exploit destructive interference, as shown in Figure 10.16. Reflecting from the first surface, the light is inverted, as is always the case when reflecting from a higher index material. Since the next interface involves a reduction in index, it will not invert the reflected light. If the thickness of the intermediate layer is a quarter of a wavelength, the two reflections will differ by half a wavelength, and interfere destructively, improving the overall transmittance.

Of course, other light frequencies will produce less of an effect, because the thickness is not a quarter of *their* wavelengths, but there will still be an overall reduction in the amount of light lost. Further reductions can be made by using additional layers tuned to other wavelengths. Typical coatings can result in less than 0.5% of incident light in the visible range being reflected, although this figure grows when the light impinges at larger angles, just as iridescent effects change. However, within the relatively narrow FOV, the effect is negligible.

Another complication results from the dispersion of light into a spectrum of colors. In Chapter 8, we saw how the effective speed of light is reduced within a material, as incident radiation drives the motion of charges localized around the atoms. These, in turn, radiate at the same frequency, influencing other charges, and the entire superposition of these waves results in slower effective propagation through the material. The response of these bound electrons is greater at higher frequencies (which is what leads to Rayleigh scattering and blue skies). In a block of glass (or a raindrop) this means that the slowing—and hence the bending—is more pronounced for violet than for red light. Refractive index isn't just one number, but changes with wavelength. This dispersion causes rainbows, which is wonderful, but it is far less pleasing when it causes *chromatic aberration*—a color-dependent blurring of the image—as depicted in Figure 10.17.

From a Great Distance: Lenses, Binoculars and Scopes • 163

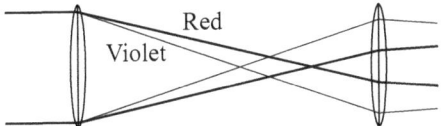

Figure 10.17 Chromatic aberration, exaggerated for clarity. Parallel rays enter a Keplerian telescope, and the amount of refraction by the objective lens on the left depends on the light's wavelength. Upon passing through the eyepiece on the right, these components will not be in focus.

The dispersion of a material can be quantified according to the *Abbe number*, V, which captures the spread of refractive indices n_R, n_Y and n_B, that are measured at specific wavelengths corresponding to red, yellow and blue, respectively:

$$V = \frac{n_Y - 1}{n_B - n_R}$$

A glass with a large Abbe number has less dispersion and causes less aberration, but unfortunately there are no materials with sufficiently low dispersion to mitigate the problem. Instead, we must replace our simple lens with several lenses with different indices. A simple version is the *achromatic doublet*, shown in Figure 10.18, where we focus on the top portion and home in on the divergence of rays for different wavelengths. The effects are exaggerated to show how the rays can be brought back to the same focus. A typical doublet will use *crown glass* with an index around 1.5 and $V = 60$ for the front, positive lens; and *flint glass* with an index of about 1.6 and $V = 40$ for the negative lens, with the two surfaces cemented together.[6]

This doublet reduces aberration, but it isn't perfect, and mid-range wavelengths will not be brought to the same focus. This can be addressed with more sophisticated approaches to forming achromatic lenses and, of course, increased cost. The eyepiece, too, must be replaced with a series of lenses to avoid color artifacts.

What started as a pair of lenses in a tube has become far more sophisticated and powerful, producing a high-fidelity image. There are a few more practical

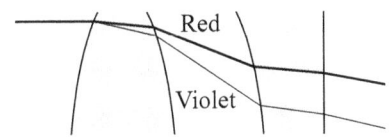

Figure 10.18 Close-up of part of an achromatic doublet. The splitting of the red and violet components by the left lens is compensated for by the second lens.

considerations that we will tie up in the next section, as well as some features of camera lenses worth exploring.

Aberrations and autofocus

Modern binoculars, telescopes and cameras are sophisticated instruments. We've touched on only the fundamentals of their operation, yet it is clear how performance trade-offs and unwanted effects needing correction lead to a plethora of designs and options. So, in this last section, we can only survey a few potential topics pertaining to optical performance.

The biggest factor driving the complexity and cost is the glass itself. Ideally, a designer could dial in the requisite refractive index and eliminate dispersion and reflectance, but real materials are not cooperative in this way. They occupy only a small (and non-optimal) part of the space of theoretically possible materials.

Some sense of this is captured by plotting the nominal refractive index versus Abbe number for various glasses, as shown in Figure 10.19. Changes in the composition and/or manufacturing produce this spread in performance. Recall that the Abbe number is inversely proportional to dispersion, so larger values correspond to less chromatic aberration. Fewer options lie this way, and they come at the price of a lower refractive index.

Optics manufacturers are loathe to share the proprietary details of their glass and its place on this diagram. They rely on monikers such as extra-low dispersion ('ED'), ultra-low dispersion ('ULD') and so on, which do not have quantitative demarcations, and are not as descriptive as one might wish.

The quest for lower-dispersion glass continues, although we've learned to mitigate chromatic aberration by using compound lenses, such as the

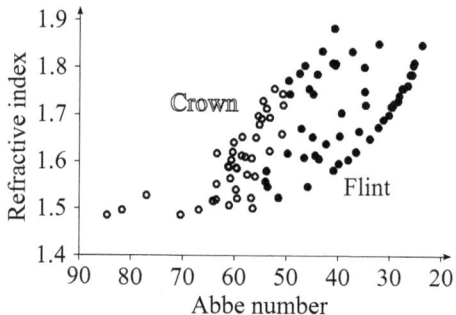

Figure 10.19 Refractive index vs. Abbe number for various types of glass, which have been broadly categorized here as crown and flint glasses. [Data from the Schott optical glass catalog.]

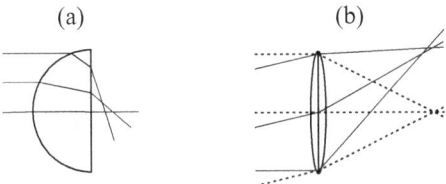

Figure 10.20 (a) Spherical aberration occurs because the focal length is different depending upon the vertical location at which light enters a spherical lens. The refracted rays do not converge at the same point. (b) Coma is a distortion that occurs off-axis, as shown by the solid lines, which do not come to a common focal point, even though light that is on-axis (dashed lines) does.

doublet of Figure 10.18. As this design cannot compensate for the mid-range wavelengths, *triplet* lenses are sometimes used. Such *apochromatic* systems perform better, but still leave other intermediate wavelengths uncorrected, and have more surfaces that lose light due to reflections and require more coatings. Less dispersion makes it easier to tune out these last vestiges of distortion. Designs are further complicated because color distortion isn't the only performance limiter that must be engineered away. Quality lenses must mitigate other contortions of the image, such as *spherical aberration* and *coma*.

Lenses are spherical because it is easy to produce them that way, not because it gives optimal performance. For such lenses, in fact, the focusing of rays to a single point, which we have been tacitly assuming all this time, never actually happens. Instead, they only come near one another: in quality optics, sufficiently near such that their spread isn't perceptible. This is *spherical aberration*, illustrated in Figure 10.20 (a). If we ask what shape we must give our optics in order to prevent this, we'll discover that a conic section (ellipse, parabola or hyperbola) must be used; but these are difficult to fashion, so such *aspherical* lenses cost more.

Another problem is *coma*, which distorts light from off-axis, as shown in diagram (b). This effect is especially apparent with stars seen through a wide FOV instrument. Those at the center will be sharp, while those around the periphery will have comet-like 'tails'. In a camera, such aberrations are reduced by stopping down the aperture, but even an astronomical telescope with a parabolic mirror (and hence no spherical aberration) will have coma.

Aberrations are not limited to the objective lens, but are a liability for the eyepiece as well. Modern instruments employ oculars that may consist of five or more elements designed to reduce disfiguring of the image. Its placement is critical too; recall that for an infinitely far-off object, the eyepiece must sit one focal length from the focal point of the objective. This produces a focused image for a normal eye in its relaxed state. Were this the only application for the scope, and if all eyes were the same, we could manufacture the device with the eyepieces fixed in location.

For closer objects, the image formed by the objective moves behind the focal point, so if the objective and eyepiece spacing isn't changed, the image will not be focused. The solution used by astronomical telescopes and many binoculars is to allow the eyepieces to be moved along the optical axis for focus control. Such 'external' focusing is easy to implement, but because it changes the length of the instrument, it is impossible to isolate the interior and exterior. That is, it isn't possible to seal such a device. Sealing not only protects the optics from accidental immersion, it prevents the more insidious condensation that may occur if air—which always contains water vapor—gets inside. Besides 'fogging' the interior surfaces, the moist environment permits mold to flourish. Quality binoculars and spotting scopes are therefore isolated from the atmosphere by filling them with nitrogen gas behind an airtight seal.

This makes adjusting the focus a challenge, because we cannot change the separation of objective and eyepiece. This is accomplished by introducing yet more glass in the form of a special focusing lens located between the objective and the prism. Changing its location has the same effect as moving the eyepiece. Rotating the focus wheel moves a shaft that slides these lenses back and forth without compromising the seal.

With a familiar pair of bins, the focus wheel motion becomes instinctive, and we learn to quickly adjust to changes in object distance. Manual focus works well enough, and of course it was for many decades the only option for photography, where turning a ring translates the position of various lens elements. But photography produces a permanent record, and it can be unforgiving if the focus isn't perfect. With experience, a skilled photographer can accurately and quickly adjust it on demand, but happily, *autofocus* (AF) technologies exist to help the rest of us.[8]

Autofocus is just one of several feedback systems cameras may employ to assess performance and make rapid compensation on the fly. Another one is image stabilization, in which motion sensors are used to make minute changes to the position of a special stabilization lens, enabling less blur and slower shutter speeds. Again, this comes at the price of introducing more optical elements with their own attendant problems.[9]

One AF strategy relies on analysis of the image in response to changing focus via contrast detection. When focus is sharp, light levels change more dramatically at edges, but when the focus is poor, the blurring washes out the contrast. This is easily quantified, and by measuring the change in contrast in an image while the focus is adjusted, the camera can home in on the lens configuration with maximum overall contrast and hence the best focus.

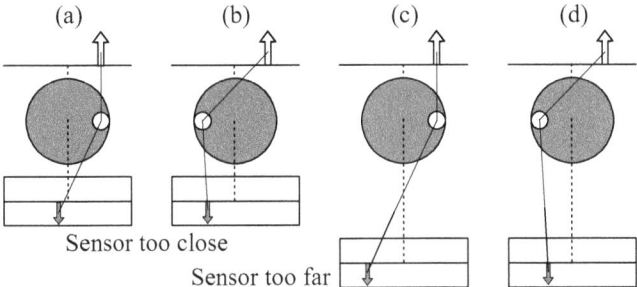

Figure 10.21 A simplified camera viewed from behind. In each diagram, the white arrow is a distant object, and the large, shaded circle is the focusing lens that is obscured except for a small region, indicated by the white circle. Light passing through this aperture leads to an image on the sensor (gray arrow). In (a) and (b), the sensor is too close to the lens, and when the constricted aperture is moved from the right (a) to the left (b), the image shifts left. Had the sensor been appropriately placed, the image would not shift at all. In (c) and (d), the sensor is too far from the lens. Now, when the constricted aperture is moved from right (c) to left (d), the image shifts *to the right*. A phase detection autofocus system samples light from different portions of the lens and exploits changes to image location to determine the optimal sensor position.

But finding this sweet spot requires sweeping through a range of poor focus positions on either side of ideal which takes time; the lack of contrast alone isn't enough to know if the focus is too short or too far.

A better method, misleadingly named *phase detection*, performs more quickly because it need not scope through a range of unfocused positions; it can determine automatically if the lenses are too close or too far and react accordingly. 'Phase' in this context has nothing to do with the wave nature of light. Rather, it refers to a shift in the *luminance patterns*: that is, the spatial offset of an image. It is easier to show this with diagrams (or by fiddling with some optics, as discussed below) than to describe in words. Figure 10.21 walks through the idea.

Each diagram shows an arrow-shaped object, a lens and a rectangular sensor array, as seen from a perspective behind the system. In each case, we contrive to block most of the light passing through the lens, except for a small circular aperture placed on one side or the other. In diagrams (a) and (b), we have placed the sensor array too close to the lens, in which case an out-of-focus image is captured, but its location will depend on which side of the lens is left open. In (c) and (d), the sensors have been moved too far away from the lens, causing the same effect; but this time the relative locations are reversed. This misregistration of images, therefore, tells us how far from the focus we are, and if we need to move the lens forward or back. By giving the camera control of the focusing motor, it can easily adjust the lens positioning and continually monitor the images to ensure that focus is kept. Doing this in a digital single-lens-reflex (DSLR) camera, which features a movable mirror as part of the shutter mechanism, requires the clever use of additional mirrors and lenses to split off portions of the light and direct them to a dedicated AF sensor.

This chapter has given us just a peek at the inner workings of modern cameras, telescopes and binoculars. These carefully engineered contrivances may not be part of the 'natural world' from an ecological point-of-view, but that demarcation means nothing from the physics perspective. And the machine in your hands that reveals the otherwise-hidden, subtle beauty of a warbler is not only no less miraculous, it warrants even more gratitude: to the physics behind it, to the generations of scientists and engineers needed to build it, and to an economic system that has made such technology feasible and affordable.

Chapter 11

Under Extremes: Heat, Cold and Thermoregulation

We seek our avian quarry under all manner of conditions and often find ourselves wondering, with great empathy, how birds can manage to survive. There's a lot going on beneath the feathers, and even in their bills, it turns out. In this chapter, we look at the effects of environmental extremes and the various strategies used by birds to mitigate their impact. We'll consider not only the regulation of body temperature but also the challenges and tactics associated with living in a watery world.

Thermodynamics and temperature

Birds have adapted to some of the most extreme conditions on earth. Except for some outrageously inaccessible locations (the floor of an ocean trench, for instance), few environments are so harsh that some species haven't found a way to live there and exploit it. Some adaptations and strategies are familiar and have been arrived at by various mammals as well. For example, it isn't surprising that penguins make use of a subcutaneous fat layer to help remain insulated from frigid Antarctic waters, or that, on sufficiently hot days, birds can be seen panting like dogs.[1] But other approaches for managing heat and cold are less obvious and occasionally surprising.

The Common Raven *Corvus corax* can cope equally well with the punishing winters of the Alaskan interior and the furnace-like heat of Death Valley, all while wearing identical 'clothes'. A downy parka, whose black surface readily absorbs the scant rays of the low Arctic sun, seems like a great idea in the far north, but not so much in a broiling desert. We'll look at how they handle this in a later section.

The management of temperature, or Thermoregulation, is of critical importance for *endotherms*.[2] This term applies to all birds and mammals and other organisms that are commonly referred to as 'warm-blooded'. It signals that the animal takes on the task of maintaining its temperature within a certain range. This is opposed to allowing the environment to (largely) determine body temperature, which is the case for *ectotherms*, or 'cold-blooded' organisms.

Our sense for maintaining a fixed temperature stems from our sensitivity to straying beyond a range of comfort. Being uncomfortable is unpleasant, but of great practical consequence, since excessive cold or heat adversely affect cell function and may cause irreversible damage. But it is worth noting that our sense of hot and cold does not always accord with the temperature. For example, if we reach into a hot oven and touch a metal pan, we will be burned immediately; but if only hot air makes contact with our skin, we can endure it briefly without incident. The metal *seems* hotter, but it has the same temperature as the air. The amount of heat energy delivered to our hand depends on additional factors, such as heat capacity and conductivity.

Still, temperature can be a useful quantitative measure of hotness or coldness, and many properties of matter are sensitive to it. By definition, temperature is that property of a system that determines if it is in thermal equilibrium with some other system. A 'system' here is some finite collection of matter that we can demarcate as separate from everything else, and thermal equilibrium is the condition in which there is no flow of heat between systems that are in contact. When two objects at different temperatures are brought together, heat will pass from the hotter to the colder object, until they reach some intermediate temperature and achieve thermal equilibrium. At that point, no more heat flows and the temperature remains unchanged.

This is all fairly clear, and beautifully underpinned by molecular and atomic physics that we are not directly aware of. Broadly speaking, a body's temperature is a measure of the kinetic energy of its constituent molecules. Two samples of gas at different temperatures have different average molecular speeds. Bring them into thermal contact, and the molecular motions in one system can jostle the molecules of the other, thereby imparting energy to them until both samples have equivalent motion, on average.

As with a gas, the molecules in a liquid have no fixed arrangement, but they are constrained to stay relatively close as they move around one another. Such a scenario is ideal for chemical reactions, and the rate at which they proceed will increase with temperature as the reactants have greater average motion and opportunity to interact.

Liquids are ubiquitous (and we are largely composed of them) so we seldom reflect on how special they are, in the context of the universe as a whole. They can only exist over a very restricted range of conditions, largely dictated by

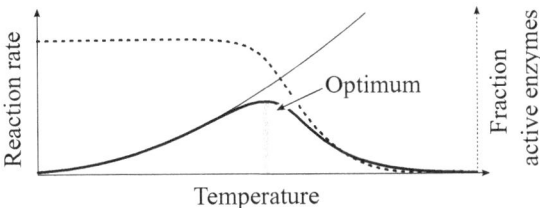

Figure 11.1 The thin solid curve shows the simple dependence of reaction rate on temperature, increasing as greater thermal energy speeds the reaction process. The dashed curve shows the effect of temperature on the fraction of active enzymes, which is flat until a sufficiently high temperature is reached and these complex molecules become denatured. The product of these two competing effects is the heavy curve, whose maximum indicates the optimal temperature for metabolic processes.

temperature. Almost all the space in the universe is far too cold—just a few degrees above absolute zero—for liquids to exist. Meanwhile, almost all matter resides in stars, where it is far too hot—on the order of many tens of thousands of degrees—for matter to be in any state other than plasma. Only a minuscule fraction of a percent of the wide range of temperatures out there are amenable for liquids or, for that matter, any interesting chemistry.

Moreover, the temperature range under which biology can proceed is even narrower. Low temperatures result in slow chemistry and slow biological processes. But sufficiently high temperatures can have a destabilizing effect, as complex biological molecules such as proteins and enzymes become *denatured* by the excess energy, losing their form and becoming fundamentally altered. Figure 11.1 shows how these competing effects lead to an optimal temperature.[3]

Since normal cell temperature typically exceeds the ambient temperature, endotherms must generate much of the heat needed to remain in an optimal state. The range of external temperatures over which body temperature is maintained simply through metabolism is the *thermoneutral zone*. Outside of this, additional measures need to be taken. If temperature is too low, an organism must reduce its thermal exposure, approach an external heat source or generate more heat internally. As an example of the first two, penguins will huddle close together in huge groups, reducing their direct exposure to the elements, while sharing scant heat that they would otherwise give up to the air.[4] Internal generation of heat via shivering illustrates the third approach, and is used by birds as well.[5]

When body temperature exceeds the usual range, excess heat must be released. One well-known strategy exploits the relationship between temperature and the motion of constituent molecules. The molecules comprising a liquid are in ceaseless motion, but have a range of speeds, and hence, energies;

and temperature is a measure of the average energy. When exposed to air, the faster molecules will be more likely to escape the liquid and enter the gas state, leaving the slower ones behind. The temperature of the liquid will be reduced as a consequence; this is *evaporation*, and this explains why it has a cooling effect. Humans have sweat glands just for this purpose, but birds (and dogs) do not. Instead, they may resort to panting, or in some cases, gular fluttering.[6] Both act to bring drier parcels of air near a liquid-holding surface in order to facilitate evaporation.

Heat management can be expressed in terms of a 'budget', with various mechanisms that can add or remove heat.[7] For the simple case in which the bird is not doing any work (such as flying), we can write the total heat gain (or loss) S as:

$$S = M - E \pm R \pm C \pm K$$

Here, M is that generated by metabolic processes, and E is the loss due to evaporation; note that M is positive and E is negative, because these processes can only add or remove heat, respectively. The other terms—R, C and K—represent the heat from radiation, conduction and convection, respectively; each of which may add or remove heat from the body. This kind of expression requires little justification. It is a simple extension of the conservation of energy, which is as easy to reckon with as balancing one's bank account based on credits and debits.

Whereas the first two terms are determined by physiological processes and certain behaviors, radiation, conduction and convection come into play at the interface between the bird's body and the environment. The best way to limit the deleterious effects of a hot or cold environment is to simply isolate from it. This is the function of *insulation*, which reduces the ability of heat to flow in or out, and is one of the primary uses of plumage. Before we consider why feathers make such a good thermal barrier—in both hot and cold environments, as demonstrated by the raven—we'll look at some surprising ways that bird anatomy is pressed into the service of thermoregulation.

Bills and legs

The variety of avian bill shapes is staggering. From the long, probing forceps of various waders, to the fine tweezer-like instruments of chickadees and warblers, to the sharp cutlery carried by raptors, it is clear that their designs are tailored to feeding. But thermodynamics can play a role here as well. The bill may serve an additional purpose: as an instrument for regulating temperature; for a dramatic example, consider the Toco Toucan *Ramphastos toco*.[8] Its enormous, iconic and extensively studied bill (the largest of all toucan species) efficiently removes excess heat from the bird's body by acting as a huge radiator. It is an ingenious co-operative use of a seemingly costly feature, with the benefit that no water loss is associated with its use, in contrast to panting or gular fluttering.

The rate at which heat can be moved from one place to another depends on many factors; but whether it is by conduction, convection or radiation, a greater area over which the transfer occurs will increase the rate. It's no surprise that the toucan's bill makes a good thermal device, and indeed it houses a network of vessels supplying blood that acts like the coolant flowing through a car's radiator. Studies have found that the flow is altered depending on the thermal conditions of the bird and its environment. When only moderate heat loss is needed, the proximal part of the bill receives blood, while more heat can be released by increasing the flow to distal regions, increasing the overall area used.

The toucan makes for a good case study, since the effect is dramatic, but there is a subtle relationship between bill area and thermoregulatory needs seen throughout the avian world. Even within a single species, bill size may change in response to different heat loss demands, as seen with Saltmarsh Sparrows *Ammospiza caudacuta* in North America.[9] Their habitat features a paucity of fresh water and shade, so avoiding dehydration while thermoregulating is critical. Since environmental conditions vary along their range, we might expect to see a positive correlation between bill size and maximum summer temperature, and studies show just that.

A larger bill may be a benefit under hot conditions, but for similar reasons may become a liability in the cold, when valuable heat is lost to the environment. This is one reason why many birds will rest or sleep with the bill tucked into their feathers, providing a layer of insulation for it.[10] When the bill sizes for multiple species are analyzed in terms of environment, clear trends are seen. A study of 214 species showed that latitude and temperature correlated to bill size generally, while species restricted to the tropics showed smaller bills with increasing altitude.[11]

All of these results can be seen as consistent with *Allen's rule*, which states the general observation that animals inhabiting colder climates have shorter and thicker limbs and appendages, which minimize heat loss by reducing surface area.[12] Animals can also reduce the ratio of their total area to volume, which helps to deal with cold because the larger volume has greater heat capacity and generative capability. This is what underlies the similar *Bergmann's rule*, which states that colder climates will tend to favor larger animals.[13] A recent study of shorebirds in Australia found persuasive evidence that these

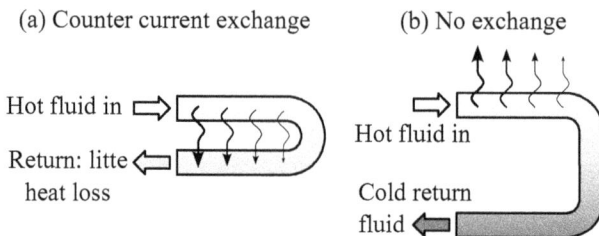

Figure 11.2 (a) Counter current heat exchange involves a hot fluid flowing into a U-shaped passage, in which there is good thermal conductivity between the two arms. As the fluid enters and cools, heat is passed back to the colder fluid, which returns with little net change in temperature. (b) A similar device in which no exchange occurs because the two arms are not in thermal contact. Heat is radiated away to the environment.

rules are driven by thermoregulatory considerations, not other root causes such as mitigating the risk of starvation or predation.[14]

Bill and limb sizes must accord with other practical constraints, and cannot be reduced indefinitely to deal with cold environments. Additional strategies to keep warm must be available. When we encounter birds in harsh winter conditions, our empathy and curiosity immediately lead us to wonder how they can manage with their exposed legs. There are competing challenges to be addressed here. Extremities must be kept sufficiently warm to avoid tissue damage; the formation of ice crystals, as occurs with frostbite, is catastrophic for cells.[15] But this requires sharing (and losing) valuable energy from the body's core, and if this is not properly managed, the bird will not be able to maintain its temperature.

The solution involves several ingenious works of engineering: mechanisms that find similar applications in other parts of bird anatomy, not to mention in animal bodies more generally (and in human-made devices). One such apparatus is the *counter current exchanger*.[16] Another key device is an *arteriovenous anastomosis*: a selectable shunt that allows for the exchanger to be used or bypassed.[17]

A schematic diagram of a counter current exchange device is shown in Figure 11.2 (a). A fluid is made to pass through a U-shaped passage, where the two arms are in proximity. The fluid cannot pass between the walls but must follow the course of the tube. However, other properties of the fluid can pass across this barrier. In our case, that property is heat.

Fluid at a high temperature enters at the top left and loses heat across the thermally conducting wall. The further the fluid proceeds on this arm, the cooler it becomes. Upon reaching the bottom of the 'U', it is at its lowest temperature. Further progress brings it into thermal contact with the hotter fluid, increasing its temperature. The end result is that the fluid is able to pass into a colder environment, yet return to the source without as much heat loss as would occur if the two arms were not in thermal contact, as shown in diagram (b).

Counter current exchange can be employed to manage more than heat; it is also useful in the context of chemical gradients for molecules that can pass through a semipermeable membrane. For the salt extraction performed within specialized glands of seabirds (Chapter 3), the flow of blood past the ducts takes a similar trajectory to make the transfer of salt more efficient. In fish gills, the same approach is used to extract as much oxygen as possible.[18]

There are other ways in which this simple idea can be used for thermoregulation. One common structure—very similar to that shown in Figure 11.2 (a)—is the *venae comitantes*, a layout in which large veins and arteries are paired up. The efficiency of exchange using smaller blood vessels is improved by increasing their contact area via a large, intertwined network, the *rete mirabile*. Both types of structures are found in birds.[19] Eyes, for example, are also 'thermal windows' that need to be carefully regulated, so exchange systems are built around these as well. While they present less area for loss than the legs, they must endure extreme forced convection conditions during flight and swimming, which will increase the rate of heat loss. The further utility of all this is that it isn't just about dealing with cold. Just as passing blood through the bill can help radiate it away, the same *rete* structures that can help protect a vulture's bare head from the cold will also let it act as a radiator when it is too hot.

Insulation and waterproofing

Most of a bird's body is thermally isolated from the environment due to the insulation provided by feathers. To insulate is to prevent or attenuate the transmission of thermal motion, so that the kinetic energy of molecules on one side of a barrier has little effect on the molecules on the other.

The best practical insulation design that humans have made is the *Dewar*, or thermos bottle, which uses a reflective inner container surrounded by a larger one, with the space between them pumped (mostly) free of air.[20] The reflective walls reduce the energy lost via radiation, while the near-vacuum provides a barrier to conduction and convection. These last two mechanisms require that energetic atoms or molecules bump into one another, which cannot readily happen if there are only a few of them about.

Vacuums are not easy to make, but the next best insulator is freely available: gas, in the form of air. As it is mostly empty space, populated by rarely interacting molecules, air cannot conduct heat effectively. The challenge is that air is unconstrained; it is a system without sufficient internal forces to make it cohere. Air needs to be corralled if we are to use it to insulate. If a solid in contact with a trapped volume of (initially cooler) air transfers heat into it, eventually the air molecules will have their energies increased, while also passing energy back into the solid, maintaining thermal equilibrium. But if that heated air isn't constrained, and cooler air comes in to replace it, the solid will continue to lose heat. Hence the cooling effect of moving air.

It is imperative that air be kept trapped if we are to utilize it for insulation. Rather than manage a single large volume of confined air, it is easier and more effective to use many small, isolated pockets. Hence, foams make for great insulation with their countless bubbles that isolate little parcels of air. Moving heat across a collection of such cavities is difficult.

Even if a structure lacks the segregated partitions of foam, it can still be a good insulator if airflow is otherwise restricted. Fiberglass insulation, seen up close, resembles the inside of a haystack: a jumble of randomly crisscrossing strands. This makes an effective insulator because the many obstacles isolate the air in more-or-less disconnected spaces.

A similar structure is created by a network of disorganized, proximal barbs of contour feathers. Lacking the order that the distal regions achieve with their interlocking barbules, the barbs closer to the skin form a mesh of randomly oriented filaments, which retain enough stiffness that the air is tightly constrained into quasi-pockets.[21] The trapped air has little means to move and so provides outstanding insulation.

A layer of downy feathers isn't sufficient protection in a world with so much liquid water, however. Wet this material and the air pockets will be depleted, greatly diminishing the insulative capability. It is critical, then, that the downy layer be kept isolated from liquids. This is accomplished beautifully by the distal part of the same feathers, which cover the downy portions of other feathers like overlapping shingles.

Still, it isn't obvious that something as (seemingly) insubstantial as a feather would make for good waterproofing. Moreover, there are multiple ways in which water and feather can interact, which will affect the nominal strategy for structuring the barbs. Consider the case of terrestrial birds that do not swim or dive, and who typically encounter water in the form of raindrops. Here, nature will favor exterior feathers that can repel water, that prevent the drops from spreading out onto the surface and allow them to bead up and simply roll off.[22]

Water forms drops because of the *cohesive* forces between water molecules, which are electrically neutral, but polar. Each molecule has a balance of positive and negative charge; but it is not arranged uniformly, so up close, the electrical field can be quite strong. Hence, they will tend to orient so that the positive part of one molecule is close to the negative portion of another, forming a bond.

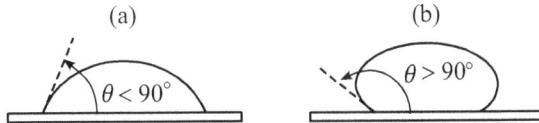

Figure 11.3 (a) A liquid droplet on a hydrophilic surface, which attracts the water molecules and results in a contact angle θ that is less than 90°. (b) A droplet on a hydrophobic surface which repels the water, leading to a contact angle larger than 90°.

Place a droplet on a solid surface and the water molecules will interact with its molecules as well. This can lead to a wide variety of results. Any attractive force between the molecules in the solid and the water is an *adhesive* force, and it is the relative magnitude of this force (between unlike molecules) compared to the water's own cohesive force (between like molecules) that determines how 'wettable' a surface is.

A *hydrophobic* material presents weak adherence forces, while a *hydrophilic* one exhibits strong attraction to water molecules. To quantify this, we need only look carefully at a droplet of water standing on the surface. The edge will meet the solid at some contact angle, as shown in Figure 11.3. If the adherence forces dominate, it will be pulled onto the surface, and the angle will be decreased, as shown in (a). If the coherence forces are greater, the drop will stay beaded up, with a large angle, as in (b).

Take away the surface, and the droplet becomes spherical, so in a sense the lack of a solid surface is the most hydrophobic surface of all. This leads to a counterintuitive way to make a surface more water-repellent: put holes (or deep grooves) in it.

The effect of surface porosity on water repellency was first presented by Cassie and Baxter in 1944 in a paper that included a discussion of duck feathers and their hydrophobic properties.[23] It introduced a simple model well-suited for feathers, consisting of an array of parallel cylinders of radius r with a distance of $2d$ between their nearest edges, as shown in Figure 11.4 (a). The contact angle was shown to depend on the adherence properties of the cylinder material as well as the size and spacing of the cylinders. By increasing the fraction of the total interface that is open, the contact angle will increase. This makes sense because the surface will always present some attractive force, but the empty spaces will not, leading to less adherence overall. An example of the relationship between the contact angle and the ratio *(d+r)/r* is shown in diagram (b).

Note that the fraction of 'contact area' that is air increases as the ratio *(d+r)/r* goes up. What the curve makes clear is that increasing the spacing between feather barbs is a 'free' way to improve its water repellency, provided the gap is smaller than the droplet size, of course.

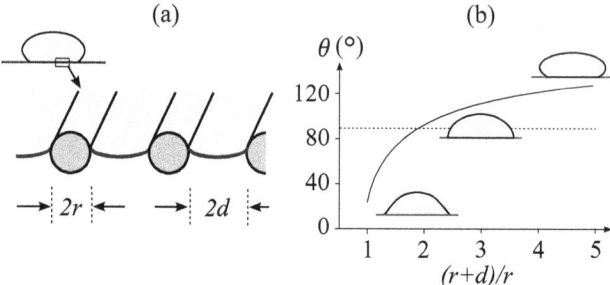

Figure 11.4 (a) The Cassie–Baxter model geometry for a droplet on a porous surface. The surface is modeled as a series of parallel cylinders shown in cross-section, with a radius r and a separation of 2d. (b) Contact angle θ versus the dimensionless parameter (r+d)/r, which quantifies the amount of porosity. When this value is 1, there are no gaps between the cylinders. The curve here is for a material that is hydrophilic, with a contact angle of 20° in this case. As the porosity increases, the contact angle eventually exceeds 90° (dashed line) and the structure has become hydrophobic.

This is fine for dealing with the drops on a duck's back, but what about the water under its belly? Here, water repellency isn't what is needed, but rather, *reduced water penetration.*[24] Repellency matters when we consider small droplets that exert negligible force against the surface. But the wall of water acting on a submersed feather exerts pressure, and if it pushes hard enough, it will pass through the gaps between the feather barbs. Our intuition suggests that large spaces between barbs may be a problem here.

A water droplet contacting a porous surface, as shown in Figure 11.4 (a), is prevented from seeping through the gaps by the surface tension resulting from the cohesive forces. The lower surfaces that 'hang down into the gaps' between the cylinders effectively behave like a strong fabric pulling up against the pressure that tries to push the liquid through.

The smaller the gap through which one tries to force a liquid, the greater the pressure needed. This is because the tighter curvature results in a greater component of surface tension directed back up into the liquid. For the 'half-droplets' hanging down in Figure 11.4 (a), which have a radius d, the pressure needed to force the liquid through will be proportional to $1/d$.

We can see the problem now: if we want maximum repellency, d should be bigger than r, keeping the ratio $(d+r)/r$ large, but the increased d will mean less pressure is needed for the liquid to pass through the gaps. Hence, we cannot maximize both effects simultaneously. Feather design must make a trade-off, depending on usage.

In a beautiful study that links feather microstructure to environmental conditions, Rijke and Jesser compared measured values for d and r across multiple species with different relationships to water.[25] They found that, for largely terrestrial birds that need water-repellent feathers, the $(d+r)/r$ ratio was large, typically in the range of seven to nine, while for aquatic birds, where preventing penetration is more important, the ratio could be as low as three. It is an elegant example of another way in which these structures can be tailored to fit specific needs.

Feathers and their functions

The variety of functions that feathers perform is astonishing, and it is difficult to think of any other invention of nature that can be put to so many different uses. A recent review of the evolution of the feather lists 26 different applications that birds can use them for, ranging from nest linings to sound generators to tools for water transportation to mechanical bracing structures.[26] Keeping this book at a moderate length prevents us from diving into the physics of many of these other novel uses.

Very little of a bird's plumage is not being utilized for multiple roles, some of them unexpected at first glance. For example, the sparsely feathered underside of the wing has been shown to play a collision-avoidance role for various species. A recent study showed that strong, contrasting underwing patterns are more likely to be present on larger rather than smaller birds, as the former have less maneuverability and presumably more to gain from visual clues that help alert them to the whereabouts of others.[27] The study also found that such patterns are more prevalent with colonial nesting species, for which collisions would be far more likely.

All other things being equal, darker (melanized) feathers would seem to be preferable. The layered, interlocking structure of melanin, which was examined in Chapter 8, acts to strengthen the keratin and make it more resistant to damage from the repetitive mechanical stresses of flight, or from abrasion. It also helps prevent damage from UV radiation and bacteria that attacks and weakens keratin.[28] The latter has been suggested as the mechanism behind *Gloger's rule* as it applies to birds; this is the observation that darker plumages are preferentially found in more humid areas, where more bacteria are present.[29]

It is well known that many birds with largely white coloration often feature dark wing tips. This has been understood as leveraging the enhanced strength of melanized keratin in an area where significant mechanical stress occurs and where feather damage would be very costly.[30] (We will see later that there may be other very good reasons for this ubiquitous upper wing pattern.) Pelicans, gannets, terns, snow geese and others show this trait, as do gulls. Research that looked at the amount of black in the distal primaries for gull species as a function of their habitat and migratory routes found that more extensive black tends to occur for species that fly longer routes, as well as those that inhabit sunnier climates.[31] The shade of a gull's mantle was found by the same study to correlate to sun exposure, likely for improved UV protection.

The use of melanin as a response to extreme sun exposure can yet seem somewhat paradoxical. After all, darker surfaces will absorb more light and generate more heat. From that perspective, birds such as ravens and vultures would seem to be ill-equipped to deal with desert environments, yet they thrive in hot and sunny locales. If the exposed parts of feathers were tight to the body and in thermal contact with the skin, this would certainly be a problem. The copious solar radiation would be directly conducted into the body. But a critical feature of plumage is something we don't see: the significant volume of space between the 'outside' of the bird and its skin. This space, full of the proximal, downy portions of the feathers that are so important for trapping air for insulation, can also be a barrier against excessive external heat flowing in.

A simple model for the interplay of incident radiant energy, feather color and environmental factors uses the idea of thermal resistance to quantify how much heat from the sun reaches the skin.[32] For dark feathers, radiation will not penetrate very far before it has been fully absorbed by the melanin. As it is the distal parts of the feather that are thus heated, their thermal energy is more readily radiated back out into the environment than through the thick downy layer toward the skin. Add to this any amount of external air movement, from wind or the bird's motion, and the environment will be an even more effective 'heat sink'. Although white feathers might seem to avoid all this by simply reflecting the light, they will yet admit plenty of radiant energy, which will penetrate deep into the plumage, resulting in heat generated closer to the skin. (As an analogy, think of how much light manages to pass through white clouds on an overcast day.) The surprising upshot is that, with sufficient thermal conductivity to the environment, darker plumage leads to less heat being transferred to the body.

It might seem that flight, with its forced movement of air against the feathers, would help to dispel heat, and that is indeed the case.[33] However, if that flight requires exertion through flapping, the benefit will be countered by the thermal cost of generating the power. Flight can be very demanding, requiring an order of magnitude increase in metabolic rate up from the basal level.[34] Moreover, the efficiency of the pectoral muscles is only in the range of 20% or so, meaning that the bulk of the energy doesn't go into thrust for flight but rather creates heat, which needs to be shed.

Managing this thermal budget associated with flight can lead to a novel demonstration of Allen's rule, which we encountered in the context of cold conditions leading to smaller appendages. The same rule can work in the opposite sense as well, as one study on wing bones showed.[35] Longer wings increase the length of vasculature and the area from which heat can be dispelled, but they are more costly in terms of aerodynamic demand. Some species in hot climates will make such a bargain, sacrificing flight efficiency and agility for improved thermoregulation.

Other examples of this trade-off were seen during studies on starlings, in which flight behavior was observed as a function of ambient temperatures.[36] Under extreme conditions, the birds would open their bills and trail their legs, both of which come at a flight efficiency cost, but which help to remove heat. The smartest way of dealing with high temperatures, which the starlings also demonstrated, is simply being less inclined to fly at all.

Thinking about thermodynamics reveals an advantage of doing the demanding work of migration at night. Not only are ambient temperatures in general lower, but there is also a significant benefit in having a dark, cloudless sky above. It acts as a huge 'sink' for the infrared radiation produced by the bird, while providing very little radiation onto it in return. Even when nighttime air temperatures should be too high for prolonged flight, as can be the case over the Sahara, birds can yet manage the strenuous migration because of the heat-absorbing effect of the clear sky above them.[37]

Perhaps the most satisfying result of the interplay between color, thermoregulation and flight is the demonstration from a recent study that the heat to be eliminated, ultimately a product of muscle inefficiency, can in turn be leveraged to improve the efficiency of flight.[38] This may occur in several different ways. One effect involves a previously unrecognized benefit of darker wings. It isn't just that they release heat more effectively to help avoid hyperthermia. Heating the surrounding air also results in a reduction in the drag force. There will be much more on air drag in the next chapter; but for now, our intuitive sense of what it means suffices. The density of air is a key factor in how easy or hard it is to move through it. Warmer air, in which the average molecular energy is higher, expands to fill a larger volume than colder air, decreasing the density and the resistance to a solid moving through it.

Another finding is that the widespread pattern of wingtips being darker than the rest of the wing may not just be about protecting this most critical portion of the primaries from UV and abrasion damage. With black wingtips releasing more heat than the proximal portions of the wing, a temperature (and hence pressure) gradient will occur in the air above the wing. This will tend to move air from the center of the bird outward, which—as we will see in the next chapter—will improve flight efficiency. With this, we come to a natural segue from the dynamics of heat to the interactions of solids and fluids moving past one another. And once again, we'll see large-scale behaviors ultimately arising from the collective effects at the molecular level.

Chapter 12

Above the Earth: Wings, Lift and Flight

The signature ability of birds to fly is so familiar that we can easily lose our childlike sense of wonder at it. But when asked to explain just how they counter gravity, we might be hard-pressed to answer. In this final chapter, we'll take on the non-trivial subject of flight, pointing out the explanatory pitfalls of oversimplification, while aiming to make the subtleties of the wing's magical work more intuitive.

Aerodynamics and Oversimplification

Birds elicit our wonder in many ways; their vibrant colors, ecological adaptations and vocal talents are a few of the examples we've already explored. But as even a small child will tell you, what defines them is their defiance of gravity. Few other vertebrates can accomplish what most bird species seem to do effortlessly.

Flight is a large topic that warrants far more space than is available here, and involves a mix of elementary and complex ideas that are fitting for something both commonplace yet seemingly miraculous. Everything we'll need comes from *classical* physics; no mysterious quantum effects or challenging-to-visualize electromagnetism are required. But despite this, many aspects of flight are not simple, and various 'explanations' for how it works are incomplete or wrong. That such accounts may turn up in popular articles perhaps isn't unexpected, but on occasion one encounters breathless reports that flight or aerodynamic lift 'isn't understood', which is entirely false.

The alleged mysteries surrounding flight result from various facile models put forth to capture it, models that are incomplete, or incorrect, and which inevitably contradict other problematic explanations. We will see examples of this as we proceed to build a picture of how wings and air interact. But first, we'll stress that the movement of air near a wing is governed by relatively concise and well-established equations, which are straightforward to *derive* from considerations of force and energy. A century of theoretical and experimental work leaves no question about their validity or practical applicability.

Where difficulty can arise is in finding *solutions* to these equations. Unlike the simple harmonic motion of a mass on a spring—where the governing equation can be solved with basic calculus—the dynamics of fluids are captured by the *Navier–Stokes equation*, which has solutions that are not amenable to being worked out on paper. Instead, software-based approximation methods (computational fluid dynamics, or CFD) must do the heavy lifting. It is in this sense that the formal mathematical solutions (especially in the case of turbulent flow) have not been fully elucidated, and that mystery persists. The Navier–Stokes equation has stumped so many bright minds that a bounty of one million US dollars has been placed on its head.[1] Search for 'Millennium Prize problems' to see what you'll you need to do to collect the winnings, and be prepared to show your work.

As for making flight more intuitive, it is ironic that the desire for clarity can lead to oversimplifications that ultimately misrepresent how it works. Even the physics education literature has been affected; a recent study looking at 135 articles found a majority of the illustrations of airflow around a wing to be incorrect.[2] With this in mind, we will proceed carefully.

To go along with the challenging mathematics and physical explanation, flight is simply *uncanny*. Perhaps this originates with our awareness of the relentless force of gravity that it manages to defeat (or the absurdity of sitting in the sky inside a 300 ton airplane). Gravity is the simplest of the forces at play in flight, the only one that won't vary in strength or direction. It acts on mass m to generate *weight* $W = mg$, where g is the acceleration due to gravity at the earth's surface (roughly 10 m/s^2). All other forces at play must come as a consequence of air molecules striking, and being moved by, the body in flight.

The simplest way in which the molecules comprising the atmosphere exert force on a body is via *buoyancy*. Our intuition is not too strained by the sight of a hot-air balloon rising, because we recognize that, as a whole, it is lighter (more specifically, *less dense*) than the surrounding air. Picture the volume of normal air that might correspond to such a balloon hanging in the sky, perhaps outlined with a dotted line. At its bottom, the atmospheric pressure must be greater than it is at the top, because the lower air bears more accumulated weight. Buoyancy is the upward force from this imbalance in pressure, which must exactly correspond to the weight of the air inside our imaginary balloon; otherwise that air would move. Replace that same volume with a real balloon full of less dense air, and that air exerts less weight, so the unbalanced buoyant force must cause it to rise. When solar heating of the ground produces a

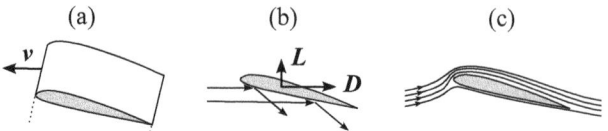

Figure 12.1 (a) A simple wing moving at velocity *v*, with the cross-section, or airfoil, highlighted. (b) From the perspective of the wing, some oncoming air is deflected off its lower surface, resulting in both backward drag *D*, and an upward force *L*, proposed by Newton as the source of lift. (c) Flowlines showing the direction that various air parcels take above and below the airfoil.

localized mass of warmer, less dense air, it will also rise, just like a hot-air balloon. This is, of course, how birds of prey utilize thermals for soaring. The birds rise because the *average* density of the bird and the warm air mass below it is still less than that of the surrounding air.[3]

The effect of rising air beneath the broad underside of a vulture's wing is obvious, but the mechanisms by which *lift* is generated by the movement of air *past* the wing are not so intuitive. And when wings are dynamically changing their shape and orientation, as is the case with flapping and hovering flight, the situation becomes quite complicated. Before dealing with these more complex scenarios, we'll consider a *fixed wing* shape cutting through the air, which is appropriate for exploring the basics of aircraft flight.

Figure 12.1 (a) highlights a defining characteristic of a wing—its two-dimensional cross-section, or *airfoil*. This shape is *cambered*; that is, it is asymmetrical, with the top surface bearing a smooth hump and the bottom being relatively flat. The size and shape of an airfoil will typically change along the length of a real wing, but we'll assume it is uniform, as shown. (We also assume the wing is very long, and will not worry about what happens at the ends.) The airfoil moves relative to the air at a velocity *v*, but going forward, we'll take the point-of-view of the wing, from which the air is moving.

We first recognize that the air will resist the motion of the wing via the frictional, dissipative force of *drag*, opposite to the body's motion. Its magnitude will depend on the density of the fluid (in this case, air), the speed of the wing and the cross-sectional area, as discussed in the context of the diving kingfisher in Chapter 5. Because the wing here is inclined relative to the airflow, the colliding air will also produce an upward force *L* (in addition to the drag *D*)—as shown in diagram (b)—an effect which Newton proposed as the cause of lift. It is a tidy explanation; but it is wrong, as it neglects what happens above the wing, and is insufficient to counter gravity.[4]

In Figure 12.1 (c), we have drawn a series of flowlines, which indicate the paths that individual parcels of air will take as they go by the wing. Note how the air passing over the top acquires a downward component as it comes off the back of the wing. Most of the air that gets accelerated downward by the wing initially lies above it, which Newton's explanation cannot account for.

Why does the air flow down the backside of the airfoil? Some authors attribute this to the *entrainment* of air due to the *Coandă effect*, which concerns how a localized jet of fluid will adhere to a nearby surface.[5] The relevance of this effect to the flow of air over a wing is questionable at best, however, and we won't invoke it.[6]

At the most basic level, we can say that the shape, orientation and motion of the wing results in some volume of the passing air being given a downward force. Newton's third law requires that the air exert an equal force upward, which, if sufficient, will counter or exceed the force of gravity. This is all correct, but it is not enough to satisfy our curiosity. We'd like to know exactly *how* the air pushes back up on the wing and understand why the air passing over the top is directed downward. Moreover, there is more to the air motion than the *downwash* left behind the trailing edge. We will work our way gently into these topics now.

Airfoils and lift

In studying the operation of the syrinx in Chapter 5, we encountered Bernoulli's principle, which explained why moving air caused soft membranes to be pulled into the passageway via a reduction in pressure. Air molecules are generally moving every which way, at speeds around 500 m/s (roughly 1,000 mph) for 'still' air. By redirecting these velocities into a preferred direction (as is the case with flowing air), they will carry less momentum perpendicularly, resulting in less force and hence less pressure. This is simply the conservation of energy as it applies to fluids, and will also impact how local air speed v and pressure P relate along a wing's surfaces. For air of density ρ, we can write the principle as:

$$P + \frac{1}{2}\rho v^2 = \text{constant}$$

Air moving past a wing is not the same as air forced down a narrow tube, but the velocity and pressure relationship must still hold. The critical effect of a wing's shape and orientation is that it causes the air flowing above the top surface to move faster than that passing below. A pressure offset must then result, with the slower air below pushing up more than the fast air above pushes down, producing lift. But in trying to explain the origin of the velocity difference in a simple way, some explanations of flight get into trouble and propagate incorrect physics.

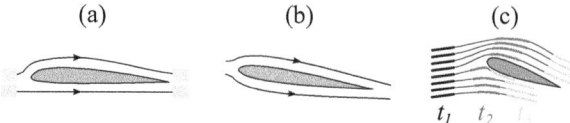

Figure 12.2 (a) The incorrect 'equal transit time' explanation of lift posits that adjacent air parcels (in gray) that pass on either side of the wing must meet up beyond the wing. This is not the case. (b) Air flow around a symmetric wing has the air parcels moving similar distances, which according to 'equal transit time' should not result in lift. But given an inclination relative to the air, as shown, this airfoil will generate lift. (c) Direct observation that 'equal transit time' is false, using localized smoke bursts when observing airflow. The heavy line sections indicate parallel flow lines ahead of the wing, at time t_1. They are seen again at a later time t_2 as they begin to flow over and under the wing, and finally in a third snapshot at time t_3, by which point the air passing over the wing has moved well past it but the air passing below is lagging well behind. [Diagram (c) modified from Babinsky (2003).[7]]

Figure 12.2 (a) shows a common—and wrong—accounting of lift. The air flowing above travels a longer distance than that below, due to the different shapes of the top and bottom wing surfaces; and there is an idea (which is false) that adjacent air parcels, shown as gray blocks, taking these different paths must 'meet up' at the trailing edge. If they are to have 'equal transit times' then the upper flow must be faster.[8]

It seems reasonable, but it isn't correct. For one thing, the offset in velocities would produce a pressure difference too feeble to enable flight. Moreover, if a *symmetrical* airfoil is sufficiently inclined—as shown in diagram (b)—it will yet acquire lift, despite the paths having equal lengths. Finally, air tunnel observations show that this 'meeting up' at the trailing edge is a myth.[9] The air that flows just over the top of the wing arrives well ahead of the air flowing below, as illustrated in Figure 12.2 (c), in which airflow speeds are captured using bursts of smoke—represented by the thicker lines—shown at three consecutive moments.

If the 'equal transit time' idea doesn't explain the increased air velocity over the wing, what does? Put simply, the differences in pressure at different points in space around the airfoil. And these pressure differences in turn result from the velocities' differences. These are reciprocal effects, and it is the attempt to avoid the chicken-and-egg nature of this relationship—and to paint a step-by-step picture of causality—that leads to oversimplified and incorrect explanations.[10]

The mutually reinforcing behavior of the velocity and pressure distributions are readily seen if we plot them for a wing in flight. Figure 12.3 (a) shows an airfoil and the velocity distribution in terms of vectors whose thicknesses correspond to the local air speed. In diagram (b) we plot the isobars—or lines of constant pressure—that result, with darker regions corresponding

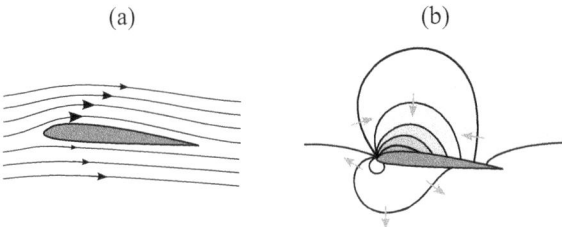

Figure 12.3 (a) Differences in velocity for flowlines around an airfoil. Larger arrows indicate higher velocities. (b) Lines of constant pressure around an airfoil. Shading indicates average pressure in a region, with lighter being higher and darker being lower pressure. The arrows here indicate the local movement that air will make in response to the pressure gradients. [Diagram (b) modified from McLean (2018).]

to lower pressure. The low pressure above the wing not only leads to lift, but also an 'upwash' in front of the wing: the pulling of air up and over, evident in diagram (a).

Thus far, we have only considered airfoils—that is, cross-sections of a wing—for simplicity. If we wish to understand the wing in three dimensions, we'll need to note that the air beyond the wingtip will also be affected by the pressure gradient and its turning effect. Figure 12.4 shows a simplified rendering of air moving in three dimensions in response to gliding wing motion. In addition to the flow over the wing, the low pressure atop the wing will create an upwash from the sides as well. That is, the difference in pressure encourages air to flow up, around the sides of the wing. Once the wing has moved past, this motion remains in the form of *vortices*: that is, rotating currents of air, which, thanks to angular momentum conservation, keep turning as they trail off on either end of the wing. Images of planes flying above a cloud bank will often capture this effect beautifully.[11]

As air curls up around the end of the wing, it reduces the pressure difference and hence reduces lift. Wing designs and modifications that mitigate this flow are therefore of great value. One common modification is the 'winglet', basically a thin vertical slab or sharp 90° upward curve, which helps impede

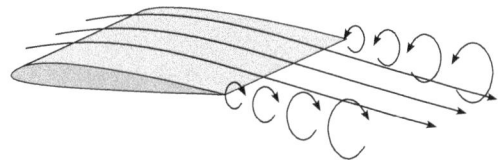

Figure 12.4 Simplified air flow over a three-dimensional wing. Air flowing over the center behaves like the two-dimensional flow that has been developed for an airfoil. At the ends of the wing, however, the pressure difference will cause a circulation to flow from below to above, causing expanding vortices, which trail either side of the wing.

air circulation. These are found on many plane wingtips, which also taper to a small width at the end, to reduce the area for the vortex to form. Many soaring birds will splay the primary feathers out to create a 'slotted' wingtip, which diminishes the vortex effect in a similar way. In the previous chapter, we saw how the thermal effects from darker wingtips, warmed by the sun, could lead to an outward flow of air along the wings, toward the tips. We can now see how this is beneficial for flight, because such a flow will oppose the circulating upwash that causes trailing vortices.[12]

Another famous way to mitigate the effects of finite wings is for multiple birds to fly in V-formation. The key here is that the distal side of a trailing vortex always forms an upwash. Each individual in a flock (except for the leader) can position itself to harvest the lifting force provided by its neighbor's upwash, and thereby expend less energy.[13]

Now that we have developed a sense for how and why aerodynamic lift occurs, we can think about quantifying it. Many factors affect the lift and drag forces on a wing, but fortunately, we need only focus on several key variables in order to build up a useful and intuitive model that will inform much of what we see in bird flight.

Airflows and flight

Newton's laws dictate that the lift generated by a wing will equal the mass of air displaced downward multiplied by the acceleration it is given. We can use this to sketch out the relationship between lift and other variables at play. First, note that the mass of air is the product of density ρ and volume, which will be proportional to the total area of the wing, A. The acceleration involves taking a parcel of air and moving it from rest up to the nominal airspeed v, and to do so over the time interval that it takes to transit the wing. That time goes inversely with velocity, so velocity change per time is proportional to the velocity squared. Therefore, the magnitude of the lift force L obeys an expression that looks just like that for drag (see Chapter 5), namely:

$$L = \frac{C_L}{2} \rho A v^2$$

where C_L is a dimensionless *lift coefficient* that depends upon various other parameters.[14]

For level flight where the bird is not accelerating up or down, the lift is equal and opposite to the weight W. Dividing both sides of the above equation by area, we then have an important expression for the ratio of weight to wing area, the *wing loading*:

$$\frac{W}{A} = \frac{C_L}{2}\rho v^2$$

When we consider most bird flight, the density of air won't change significantly, and provided that the coefficient is constant, we see that the weight supported by a given wing size will vary with the velocity squared. Heavier birds need to fly faster to stay aloft. Conversely, for birds with similar mass but different wing shapes, a larger wing area allows the bird to maintain lift at lower speeds.

The weights and wing dimensions for flying birds cannot vary independently; big birds with large wings will be heavy. If a length d characterizes the size of a bird, then its weight will be proportional to the volume, or d^3, while its area scales as d^2. Wing loading will therefore increase as d increases. Stated differently, wing loading cubed should be proportional to weight, and indeed this trend holds on average, as shown in Figure 12.5, which plots results for a variety of species using logarithmic axes. The line has a slope of 3, as it should for a cubed relationship. If we extended this plot to include insects as well as our own flying machines, up to and including our largest aircraft, we'll find that the same trend holds.[15]

The coefficient C_L is obviously key here, and it depends on several factors, including: the viscosity of air; the details of the airfoil shape; and the orientation of the wing relative to the flow, the so-called *angle of attack*, α. This last factor is especially important and warrants a closer look, because it is easily and often changed during flight.

Figure 12.6 illustrates the angle of attack for various airfoil shapes. A positive angle is what makes it possible for a symmetric airfoil (as shown in diagrams (a) and (b)) to generate lift, because it breaks the symmetry of the upper and lower surfaces, as seen by the incoming airflow. The difference in α for these cases makes it clear that the higher angle will increase the downward force on

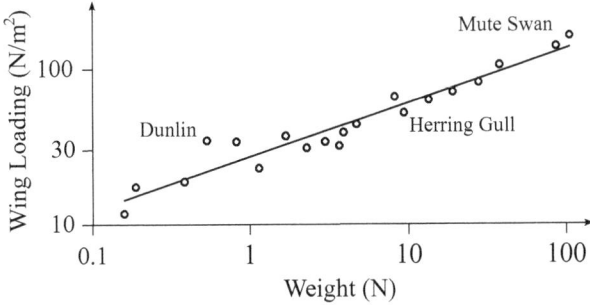

Figure 12.5 Wing loading, defined as weight per cross-sectional area of the wings, versus weight (on a logarithmic axis) for a variety of birds. The line has a slope of 3, indicating that wing loading cubed is proportional to weight. [Data from Tennekes (2009).]

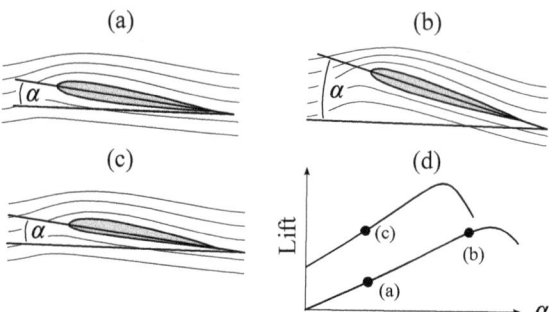

Figure 12.6 (a) A symmetric airfoil with a small angle of attack α. (b) The same airfoil with a larger angle of attack. (c) A cambered (asymmetric) airfoil with the same angle of attack as in (a). (d) Lift as a function of angle of attack for the two airfoils. Although cases (a) and (c) have the same α, the cambered airfoil has more lift. For the symmetric airfoil to generate comparable lift, it must take a higher α as in (b). In both cases, a maximum angle is reached, beyond which an increased angle reduces the lift, which is the stall condition.

the air. In diagram (c) we have a cambered (asymmetric) airfoil that requires a lower angle of attack. In (d) we plot lift versus α for both shapes, showing that it increases with angle for a while before reaching a maximum: the condition of *stall*. So long as we stay below this critical angle, the coefficient C_L will be proportional to α.

The behavior of the airflow around the wing as α increases is worth describing. We first must recognize that the kinds of streamline flows in the diagrams so far in this chapter have not shown what happens near the airfoil. Due to the roughness inherent in any surface, the air molecules will effectively adhere to it, creating a 'no-slip' condition of stationary air. Further from the surface, the velocity will gradually get larger, until it reaches the value typically indicated by the nearest streamlines in the diagram. This boundary layer has important effects, as it impacts the amount of drag experienced by the wing, as well as the lift.

At the leading edge of the wing, the flow of air above the boundary layer will usually be laminar; that is, it moves along a set of smooth, parallel flow lines. The pressure gradient that develops atop the wing pulls the oncoming air up and over; but at the trailing edge, the situation isn't so simple, because the pressure must eventually go back up. The result of this is illustrated in Figure 12.7, which shows flow over an airfoil at increasing α. The air sliding over the back of the wing encounters an adverse pressure gradient that opposes the flow, and the competing forces will cause it to develop a *turbulent* pattern, in which the flow is no longer smooth.

As α increases, the pressure at the front rises, but so must the gradient. Higher attack causes progressively more turbulent flow that grows in extent

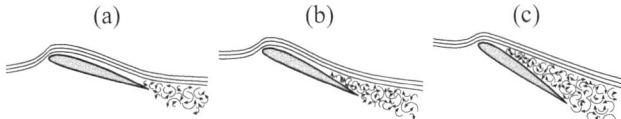

Figure 12.7 (a) At a modest angle of attack, the airflow over the airfoil is mostly laminar except for a small turbulent layer that has separated at the back of the wing. (b) Increasing the angle of attack moves the separation layer further up the airfoil. (c) Near the stall condition with the very high angle of attack, the turbulent layer extends over most of the upper section of the airfoil.

and moves 'up' the airfoil, increasing the drag. Once the angle has increased up to the stall condition, the turbulent flow has reached all the way up the airfoil, the boundary layer has fully separated off the wing, the drag is substantial and the lift stops increasing.

The limitation on α that is imposed by the stall condition obviously restricts a plane or bird from harvesting more and more lift via the relatively simple procedure of increasing the inclination of the wing. There are ways to overcome this—to a degree—such as using an additional surface near the leading edge of the wing to 'prepare' a thin, fast flow of air that will help restore a favorable pressure gradient back down the wing. Such is the function of the slats deployed on the leading edge of a plane on takeoff and landing, appearing like miniature wings with a small gap between them and the wing proper. For birds, the *alula* feathers that attach at the 'thumb' perform the same role when they are lifted, creating a slot through which air is funneled over the wing surface.[16]

Stalling is of practical value when used in a controlled manner to bring flight to a stop, and it is particularly easy to observe in larger birds as they alight. Herons and egrets are valuable to observe as they come in to land, especially if the upper wing surface is visible. Often, the extensive turbulent flow associated with the stall will be seen to clearly rustle and displace the wing coverts.[17]

The drag force is a boon for landing, but otherwise it is a liability, as it robs forward momentum and necessitates that we overcome it with another force: *thrust*. Aircraft obtain this force from jet engines or propellers, while birds generate it in the far more complex motions of flapping, which we will get to in the final section. For now, we'll just note that thrust is obtained at the cost of energy expenditure, and during steady flight, the thrust will cancel out the drag force.

We encountered drag in the context of a diving kingfisher in Chapter 5, where we saw how resistance to movement increased with the square of the velocity; the equation is identical in form to that for lift, developed above. The

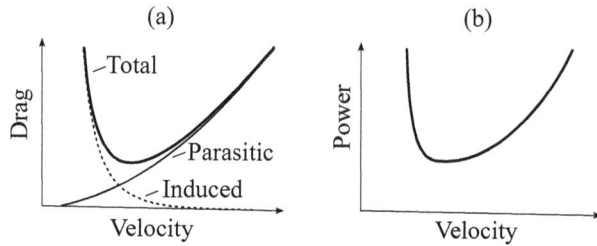

Figure 12.8 (a) Drag vs. velocity involves both parasitic drag (thin solid line) and induced drag (dashed line), which when added yield the total drag (heavy solid line). (b) Power vs. velocity shows a similar structure, as the required power is the drag times the velocity. There is a clear minimum indicating a velocity that requires the least power, while slow and fast flying involve larger energetic demands.

movement of any object against a fluid will always provoke such resistance, sometimes termed *parasitic drag*. Not only is this ever-present during flight, there is also an additional drag that occurs whenever a three-dimensional, finite-length wing moves through the air and harvests lift; this is known as *induced drag*.[18]

Induced drag is a consequence of the downward flow of air that the wing creates. In level flight, the movement of air relative to the wing is not exactly opposite the direction of motion, but has a downward component. This means that the aerodynamic lift force, which is at 90° to the airflow direction, is not purely vertical, but 'leans back'. It has a large vertical component that counters the force of gravity, but also a component directed backward—this is the induced drag. For a wing at high airspeed, the downwash velocity is small in comparison, so the effect is far less pronounced than when flying slowly. In that case, the downward flow becomes a much larger component, and the situation becomes something like flying in a downward-directed wind. Induced drag is found to vary with the airspeed v as $1/v^2$.

Plotting the total drag force versus velocity, as shown in Figure 12.8 (a), we can see the competing trends from parasitic and induced drag result in a clear minimum. We can go a step further and calculate the *power*: the rate at which energy is used. Steady flight requires a thrust force T equal and opposite to the drag D; and in moving a horizontal distance Δx, the energy required is $T\Delta x$. Since power P is the energy use per time, we divide this quantity by Δt. Since $\Delta x/\Delta t$ is the average velocity, we have $P = Dv$, and plotting this in Figure 12.8 (b) reveals a clear minimum in power versus velocity.

The implication, then, is that for a given wing shape and orientation, which will determine the drag characteristics, there will be an optimal speed that keeps the energy demands on the bird to a minimum, and that flying very fast and very slow are especially demanding activities.

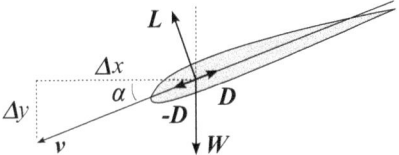

Figure 12.9 In gliding flight, the airfoil is directed downward at an angle α and moves at a constant velocity **v**. The drag force **D** is opposite to this direction, and is balanced by a component of the weight **W**, denoted **-D**. The lift **L** balances the remainder of the weight, leaving zero net force on the airfoil, which moves at constant velocity, covering a horizontal distance Δx while dropping Δy.

Before looking at how thrust is developed by flapping, we finally note that if no thrust is supplied (and if no lift is available due to air currents or thermals), the only motion possible is *gliding* flight. In that case, the descent is characterized by a small angle α with the horizontal, as shown in Figure 12.9. The weight of the bird *W* has a small component along this direction (*-D*), which acts to cancel out the drag *D*, which is directed opposite to the motion. Since there is some airflow over the wing, there is still lift *L* being generated, but it isn't sufficient to keep the bird aloft.

Inspecting the geometry of Figure 12.9 reveals that the ratios L/D and $\Delta x/\Delta y$ are the same. This is the *glide ratio* and is a measure of how far a bird or plane can travel as it drops a given distance. Among birds, the Snowy Albatross *Diomedea exulans* has been identified as a top glider, with a glide ratio of 25—twice that of a Common Tern *Sterna hirundo*—while sailplanes have glide ratios ranging up to 60.[19]

Flapping and hovering

To generate forward thrust, birds must exert a force that directs air backward. Since wing motions are often very rapid, it can be difficult to see exactly how they manage to do this. Although flapping might appear a simple up-and-down movement, it is in fact much more involved, with significant changes to the shape and orientation of the wing during the cycle of a wingbeat.

The complete course of a typical wingbeat can be roughly divided into two parts: the down, or *power stroke*; and the up, or *restoring stroke*. The downstroke gets the 'power' moniker from the fact that the bulk of the aerodynamic force is generated during this period. This accords with the fact that, for almost all birds, the pectoral muscles that attach at the central keel and pull the wing down are significantly larger—by a factor of five or more—than the muscles that lift the wing back up.[20]

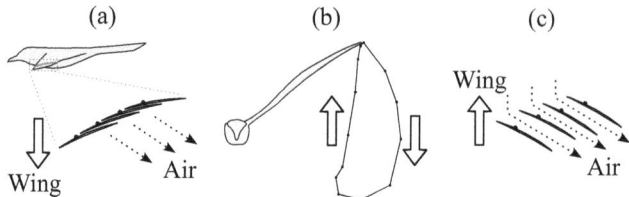

Figure 12.10 (a) Schematic showing the cross-section of several primary feathers during the downstroke. The feathers lie flat against one another, and air is forced downward and backward. (b) The motion of a wingtip as seen from ahead of a flying bird. The wing is shown in its highest position, before starting the downstroke, which is nearly semicircular. At the bottom, the wing becomes folded somewhat in to the body, reducing its cross-sectional area during the upstroke. (c) Schematic showing the cross-section of several primary feathers during the upstroke. Each feather has some ability to rotate about the rachis, and has an asymmetric vane. The downward force of the air will readily cause the vanes to separate and allow the air to pass through, reducing resistance. [Diagrams (a) and (c) modified from Dvořák (2016).[21] Diagram (b) modified from Hainsworth (1988).]

Typical flapping flight uses the primaries, which attach to the 'hand' bones, and the secondaries, which attach along the forearm, in different ways. The role of the secondaries is by far the simpler; because there isn't much range of vertical motion here, this portion of the wing effectively behaves like the fixed wing of an airplane, and has a cambered profile that produces a steady lifting force from the oncoming airflow.

The motion of the primaries is more complicated due to movements at the elbow and wrist, which augment the simple up-and-down pattern caused by the breast muscles. During the downstroke, the leading edge of the outer wing is turned so that it lies below the trailing edge, as shown in Figure 12.10 (a). In this figure, we show a cross-section of the primary feathers, which lie flat—one against the other—forming a strong and uniform surface. As the downstroke proceeds, the air below is forced both downward and backward, generating forward thrust. Meanwhile, the elbow and wrist are extended, maximizing the wing's area and reach, as is evident from diagram (b), which shows the course of the wingtip as seen from in front of the bird throughout a cycle. During the downstroke, the tip sweeps out a large, fairly semicircular path.

If the upstroke were simply a reversal of the downstroke, it would effectively cancel out its effects. So the bird must return the wing to its highest position, to prepare for the next downstroke, but without pushing much air up and forward. By bending at the elbow and wrist, the wing is pulled inward and its overall surface area is reduced, presenting a smaller cross-sectional profile for the upstroke, as can be seen by the wingtip's path—traced out in diagram (b)—which is much closer to the body. In comparing (a) and (c), we see the reason for the curious asymmetry of primary feathers.[22] The extent of the vane on the trailing edge is larger than that on the leading edge; this means that, when the wing is being lifted, the air flow will preferentially push on the larger area and force the feathers to rotate about their shafts, thereby opening up spaces between them, like louvers. (Something similar occurs in the sport of rowing; while submersed, the flat 'spoon' of the oar is perpendicular to the water, but during the recovery stroke, the lifted oar is rotated to lie parallel to the water, cutting more effectively through the air.)

Starting from rest will usually require a significant expenditure of flapping energy, since the power requirements are high at low speeds. Once in level flight and the optimal speed has been reached, less effort is required on the downstrokes. The situation is quite similar to the use of skis, skates or rollerblades. In starting from rest, one must push off with significant force, at an angle to the intended direction of motion. Once the desired forward velocity is obtained, only modest force needs to be used for the sole purpose of countering the frictional forces that would otherwise eventually bring one to rest.

Although the motion is too rapid to discern with our eyes alone, it is immediately clear that hummingbirds use their wings in a manner very different from other birds. Whereas some birds can manage a few moments of hovering, often with assistance from wind, almost everything about the mechanics of hummingbird flight is different.

The surface of the hummingbird wing is predominantly formed by the primaries—they can comprise over 75% of the total wing area, as opposed to roughly 50% for most other birds.[23] It isn't surprising, then, to find that the bones corresponding to the 'hand' of the bird are greatly enlarged, while the rest of the arm is quite small. What is startling is that the wrist and elbow joints are largely fixed in place, while the shoulder has an enormous range of motion. The ensuing cycle of motion is so different from the usual flapping that what would be called the downstroke is more of a forward motion, while the upstroke mostly moves back, as shown in Figure 12.11.

As seen in the figure, the wing first sweeps forward such that its upper surface remains 'on top' so that the underwing will strike the air, directing it down and forward. But during the reverse stroke, the wing is twisted such that the upper wing lies below the leading edge, comprising the surface that forces air down and back. Viewing from above the bird, we'll see the underside of the wing.

The forward and reverse motions given to the air during the two phases of the cycle quite clearly must impart equal and opposite horizontal forces, leaving only a net downward lift that counters the bird's weight. But if the bird inclines the plane of this pattern, the lift force will no longer be vertical, but

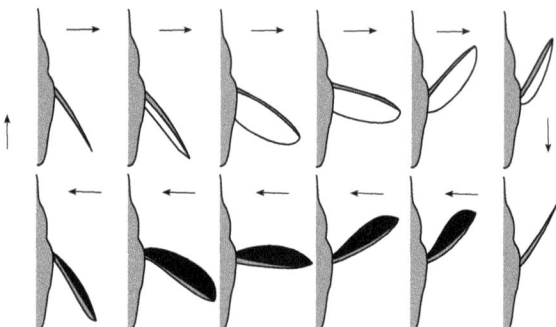

Figure 12.11 Hummingbird stroke pattern as seen from above. At the top left, the wing is extended fully back and about to begin the forward stroke. The top side of the wing is colored white, and soon becomes visible as the wing is swept forward. At the bottom right, the wing has reached the forwardmost position and begins to sweep back. The underside of the wing, colored black, now becomes visible because the top surface of the wing is now below. [Modified from Greenewalt (1991).]

will drive the bird forward (or backward), allowing it to travel while remaining suspended.

Although a hovering bird has no net velocity on average, the speed of the wings is obviously significant. We can account for it in terms of the arc swept about by the tips and the rate at which this occurs, which we denote ω. (A similar rate is useful to describe flapping flight, which we will return to shortly.) The speed of the wing tip is proportional to this rate and to the length of the wing.

Obtaining a high tip speed would seem to require that the muscles contract and relax at an extreme rate, and they do. But there are physiological limitations to this, and above some nominal 'muscle strain rate' there will be a loss in power.[24] Hummingbirds get around this with significant alterations in their bones. While there is no flexion at the elbow or wrist, the humerus rotates about its long axis in tandem with the motion at the shoulder, in order to leverage the same change in muscle length into a larger arc swept out by the wing. A much larger angular velocity can then be obtained without the muscles having to change any faster than the physiological optimum.

Dividing the angular frequency by muscle strain rate gives a measure of the mechanical leverage, or *transmission ratio*; and it isn't surprising that this value is higher for hummingbirds (around 3.0) than for other bird families (more in the range of 2.5). Smaller fliers—insects—must use an even higher transmission ratio; and the hummingbird fits well along a broad trend linking this metric to the animal's mass.

That wing frequency and bird size are related becomes apparent with even a little experience observing birds that use flapping flight. We expect that

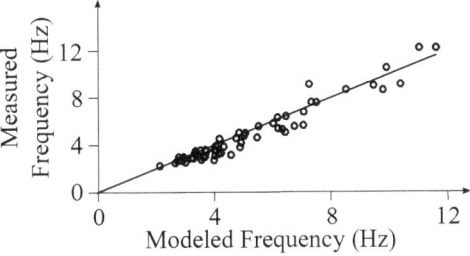

Figure 12.12 Measured vs. predicted wingbeat frequency, based on the initial model from Pennycuick (1990). The line passes through the origin and the value of β was chosen to yield a slope of 1.0. [Data drawn from Pennycuick (1990, 1996, 2001).[25]]

mass, wingspan and total wing area should all play a role; and rather than churn through a derivation, we can look to careful analyses of observed flight behavior in order to construct a model. A study using field data for 32 morphologically distinct species determined the following empirical relationship:[26]

$$f = \frac{\beta m^{\frac{1}{3}}}{b S^{\frac{1}{4}}}$$

Here, f is frequency, m is mass, b is wingspan and S is wing surface area. The β term depends on the acceleration of gravity and the air density, which we'll assume is fixed here. We see that larger mass requires an increase in flapping rate, but the effects of increased wingspan and surface area together have a larger effect that significantly reduces the frequency. Figure 12.12 shows just how accurate this simple model is. Nominal mass and wing dimensions were used to first calculate expected frequencies for various species. These results are then compared to real-world results by plotting the measured wingbeat frequencies versus the model. The data clusters tightly around the one-to-one line, indicating the usefulness and accuracy of this model.

A feel for such relationships is, of course, a valuable birding tool, another reason why even the most common of birds deserve careful attention. After years of watching ubiquitous Mallards *Anas platyrhynchos* in flight, for example, and having a good feel for their wing frequency, one is more easily alerted to different duck species at a distance too far for field marks or size judgment, simply because their flapping rates are noticeably different.

An entire book could be devoted to flight alone (and many have). In choosing to address features from the wider worlds of birds and birding—and nature more generally—only the briefest survey of aerodynamics is feasible. Hopefully, the familiar but uncanny nature of flight is more wonderful, and more beautiful, after having peered behind the curtain for a spell. With one last bird, and its unique method to exploit fluid flow, we'll wrap up our outing.

Coda: Looking Back

Phalaropes

I might have written about an encounter with some Wilson's Phalaropes *Phalaropus tricolor*, and had sufficient material on which to base most of this book. Liable to pop into view when training one's optics on a distant wetland, these photogenic birds sport blacks, whites, reds and even subtle blue-grays in their plumage, showcasing various mechanisms of color production in feathers. Their life rhythms are driven by the axial tilt of our planet, making them long-distance, equator-crossing migrants. And when they arrive *en masse*, they are wont to form murmurations no less spectacular than those of starlings.[1] We think of them as 'shorebirds' but they are better classified as marine birds; they possess salt glands like those of an albatross.[2] This allows them to exploit punishing environments like the Great Salt Lake, a place teeming with tiny invertebrate organisms but without fish to feed on them, providing ample food for enterprising birds.[3] Female phalaropes lay their beautifully patterned, pyriform eggs and immediately head south, leaving the males to incubate them.[4] Their call is a lovely nasal beeping, rich in harmonic content.

Picture a phalarope floating on shallow, clear waters, its long legs terminating in strange, lobed feet that are just visible beneath it, appearing shorter due to the refraction of light. It's hard to imagine it sitting still. These birds are, after all, inseparable from their signature behavior: a unique proclivity to spin about in place, instigating waves that radiate out in every direction. Their rotational motion would continue indefinitely, thanks to the conservation of angular momentum, but the drag forces exerted by the water prevent it. So the bird maintains its motion by kicking backward, turning one way as the water beneath is forced into a vortex flow in the opposite direction. And that is the point of this effort: not for the bird to spin, *per se*, but to circulate the water.[5] The phalarope's dizzying gyration is an artifact, an equal-and-opposite consequence. It is the turning of the deeper waters that makes this little bird's strategy so ingenious.

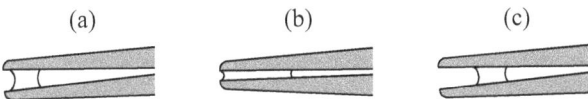

Figure C.1 Schematic of one cycle of the capillary ratchet. (a) The droplet begins near the tip with the bill open, adhering to both surfaces. (b) The bill is partially closed, squeezing the droplet, which spreads further to the right due to the proximity of the surfaces. (c) Re-opening the bill, the droplet tightens up from cohesion, causing the left side of the drop to move further inward, resulting in a net shift away from the tip. [Modified from Prakash, Quéré and Bush (2008).]

The non-zero velocity of the circulating water must come with a reduction in pressure, just as with air moving over a wing or through a syrinx. And the adjacent water responds, pulled across the gradient, into the flow. If the lake bottom isn't too far below the kicking feet, the water there, and its living, edible content, will be sucked up through the vortex and welled up to the surface. The spinning is all about getting at food that is normally beyond the bill's reach.

Many of the morsels may be microscopic in size, so the bird's good visual acuity is key, as is the long, straight, tweezer-like bill. But moving a food-rich droplet from the bill's tip up to the mouth presents a challenge. Phalaropes don't tilt the head back like a toucan wolfing down a chunk of banana. Instead, they exploit the adhesive and cohesive properties of water—relating to the same physics that determine how it beads up on feathers—together with a clever bill motion known as the *capillary ratchet*.[6]

The ratchet action is captured in Figure C.1. Diagram (a) shows the bill in an open position with a droplet near the tip. The water adheres to each mandible, but because of the angle between them, water in the droplet skews right where the surfaces are closer and the attraction greater. When the bill is partially closed, as in (b), the squeezed droplet expands, but preferentially to the right, as the surfaces are nearer to each other. Upon opening the bill, as shown in (c), both sides of the droplet retreat inward, with the cohesion of the water pulling the droplet back together; but again the symmetry is broken, and the molecules want to move rightward toward the greatest adhesive force. As a result, the drop as a whole has shifted away from the tip, and with several more identical cycles, it will be transported to the mouth.

Watch your next phalarope carefully and, if it's feeding, you'll see that what otherwise might have appeared to be an unremarkable 'chewing' motion is instead this process, repeated over and over. As with the unseen vortices turning beneath the bird, it's another subtlety that can easily go unnoticed over a lifetime of birding. But once glimpsed and understood, it's the kind of peek behind the curtain that can get you wondering what other physics might be at play.

'BVD' and dipping

During one South American birding excursion, at the end of the first day our guide asked if we had any 'BVDs' on the day's list. It meant '*better view desired*'. Were there any species that we'd seen (or heard) just well enough so that they could be ticked, but which provided a too-brief and/or underwhelming encounter? Something we should stop and take in more fully if we encountered another? I thought it was a marvelous thing to ask, and I've been preaching the acronym and the concept with guides ever since.

Aside from bad luck and the uncooperative nature of many birds themselves, we may be left desiring better views because we're trying to cover so much ground, to move from target to target, to maximize the day's list. This is no less the case in trying to tick lots of boxes in our survey of the physics that underlies birds and birding. I understand that this book will leave every reader with BVDs in that regard. As with optics, it's all about the trade-offs. The goal was to highlight the bigger picture by painting a number of smaller ones, thumbnails that hint at the taxonomy of physics and how it relates to our love of birds.

Where an explanation here has shed some light, but left only a fleeting image, I hope the reader will be inclined to come back out into the field again. The Notes and Bibliography that follow point to where to find a few guides. I have aimed, where possible, to cite literature that is available online and not behind a paywall.

Of course, at the end of any outing, we'll also ruminate on the inevitable 'dips': that is, target species that we simply didn't encounter. There are too many interesting intersections of physics and ornithology to cover here, and I regret that more could not be included, but 12 chapters of anything is more than enough. A few of the dips include the silence of an owl's wings; the production of sounds from specialized feathers (such as with snipes, nighthawks and some manakins, for example); why hummingbirds hum; the origins of color on skin, legs, bills and eggs; avian echolocation; the structural physics of nests; the energy budgets of migrating birds; the bracing structure of woodpecker tails; how hummingbirds drink; underwater swimming as another example of flight; and more features of modern binocular and camera technology, such as image stabilization. The physics of all these things are no less fascinating than the birds themselves, and they await being unraveled and enjoyed.

'The world will never starve for want of wonders; but only for want of wonder.' —G.K. Chesterton

Notes

Chapter 1 At the Feeder: Birds, Mathematics and Symmetry

1   Ditz, H.M. and Nieder, A. (2016) Numerosity representations in crows obey the Weber-Fechner law. *Proceedings of the Royal Society B* 283: 20160083. https://doi.org/10.1098/rspb.2016.0083
2   Pepperberg, I.M. (2006) Grey parrot (*Psittacus erithacus*) numerical abilities: addition and further experiments on a zero-like concept. *Journal of Comparative Psychology* 120 (1): 1–11. https://doi.org/10.1037/0735-7036.120.1.1; Pepperberg, I.M. (2002) *The Alex Studies: Cognitive and Communicative Abilities of Grey Parrots*, 1st edition. Cambridge, MA: Harvard University Press; Kirschhock, M.E., Ditz, M.H. and Nieder, A. (2021) Behavioral and neuronal representation of numerosity zero in the crow. *Journal of Neuroscience* 41 (22): 4889–96. https://doi.org/10.1523/JNEUROSCI.0090-21.2021
3   Kirschhock, Ditz and Nieder (2021).
4   Rugani, R., Fontanari, L., Simoni, E. *et al.* (2009) Arithmetic in newborn chicks. *Proceedings of the Royal Society B: Biological Sciences* 276 (1666): 2451–60. https://doi.org/10.1098/rspb.2009.0044
5   Wigner, E.P. (1969) The unreasonable effectiveness of mathematics in the natural sciences. *Communications on Pure and Applied Mathematics* 13 (1): 1–14.
6   Tegmark, M. (2015) *Our Mathematical Universe: My Quest for the Ultimate Nature of Reality*. New York, NY: Vintage.
7   eBird (2021) eBird: an online database of bird distribution and abundance [web application]. eBird, Cornell Lab of Ornithology, Ithaca, New York. Available at: http://www.ebird.org (accessed: February 21, 2024).
8   Apostol, T. and Mnatsakanian, M. (2007) Unwrapping curves from cylinders and cones. *American Mathematical Monthly* 114 (5): 388–416. https://doi.org/10.1080/00029890.2007.11920429
9   Narushin, V.G., Romanov, M.N. and Griffin, D.K. (2021) Egg and math: introducing a universal formula for egg shape. *Annals of the New York Academy of Sciences* 1505: 169–77. https://doi.org/10.1111/nyas.14680
10 Stoddard, M., Yong, E.H., Akkaynak, D. *et al.* (2017) Avian egg shape: form, function, and evolution. *Science* 356 (6344): 1249–54. https://doi.org/10.1126/science.aaj1945
11 Narushin, V.G., Griffin, A.W., Romanov, M.N. *et al.* (2022) Measurement of the neutral axis in avian eggshells reveals which species conform to the golden ratio. *Annals of the New York Academy of Sciences* 1517 (1): 143–53. https://doi.org/10.1111/nyas.14895
12 Narushin, Romanov and Griffin (2021).
13 Stoddard, Yong, Akkaynak *et al.* (2017).

14 Barnsley, M.F. (1998) *Fractals Everywhere*, 2nd edition. Boston, MA: Academic Press Professional.
15 Mandelbrot, B.B. (1979) Discussion paper: fractals, attractors, and the fractal dimension. *Annals of the New York Academy of Sciences* 316: 463–4. https://doi.org/10.1111/j.1749-6632.1979.tb29489.x
16 Tucker, V.A. (2000) The deep fovea, sideways vision and spiral flight paths in raptors. *Journal of Experimental Biology* 203 (24): 3745–54. https://doi.org/10.1242/jeb.203.24.3745; Tucker, V.A., Tucker, A.E., Akers, K. *et al.* (2000) Curved flight paths and sideways vision in peregrine falcons (*Falco peregrinus*). *Journal of Experimental Biology* 203 (24): 3755–63. https://doi.org/10.1242/jeb.203.24.3755
17 Datseris, G., Kottlarz, I., Braun, A.P. *et al.* (2023) Estimating fractal dimensions: a comparative review and open source implementations. *Chaos* 33 (10): 102101. https://doi.org/10.1063/5.0160394; Aks, D.J. and Sprott, J.C. (1996) Quantifying aesthetic preference for chaotic patterns. *Empirical Studies of the Arts* 14 (1): 1–16. https://doi.org/10.2190/6V31-7M9R-T9L5-CDG9
18 Pérez-Rodríguez, L., Jovani, R. and Mougeot, F. (2013) Fractal geometry of a complex plumage trait reveals bird's quality. *Proceedings of the Royal Society B* 280: 20122783. https://doi.org/10.1098/rspb.2012.2783
19 Quigg, C. (2019) Colloquium: a century of Noether's Theorem. *arXiv: [physics.hist-ph]*. https://doi.org/10.48550/arXiv.1902.01989
20 Merriam-Webster (2024). Available at: https://www.merriam-webster.com (accessed July 25, 2024).
21 Edelaar, P., Postma, E., Knops, P. *et al.* (2005) No support of a genetic basis of mandible crossing direction in crossbills (*Loxia* spp.). *The Auk* 122 (4): 1123–29. https://doi.org/10.1093/auk/122.4.1123
22 Conklin, J.R., Verkuil, Y.I., Riegen, A.C. *et al.* (2019) How wry is a wrybill? *Wader Study* 126 (3): 228–35. https://doi.org/10.18194/ws.00164
23 Magat, M. and Brown, C. (2009) Laterality enhances cognition in Australian parrots. *Proceedings of the Royal Society B* 276: 4155–62. https://doi.org/10.1098/rspb.2009.1397
24 Three of the four fundamental forces—namely, electromagnetism, and the weak and strong nuclear forces—are consequences of *gauge symmetries*. Many pages would be needed to do the subject justice here. For a gentle introduction, see Schumm, B.A. (2004) *Deep Down Things: The Breathtaking Beauty of Particle Physics*. Baltimore, MD: Johns Hopkins University Press.
25 For a short but profound hymn to mathematics that should cause even the most math-averse to seek out their old algebra textbook, look to Lockhart, P. (2009) *A Mathematician's Lament*. New York, NY: Bellevue Literary Press.

Chapter 2 In the Garden: Hummingbirds, Flowers and Forces

1 The average male comes in right at 2.5 g, per Johnsgard, P.A. (1983) *The Hummingbirds of North America*. Washington, DC: Smithsonian Institution Press. This is also the weight specified for the penny by the United States Mint.
2 His tomb at Westminster Abbey declares: 'Mortals rejoice that there has existed so great an ornament of the human race' and few physicists consider this sufficient praise for the difficult man that founded the entire discipline.

3   Henry, J. (2017) Newton and action at a distance. In Schliesser, E. and Smeenk, C. (eds) *The Oxford Handbook of Newton*. Oxford: Oxford University Press. https://doi.org/10.1093/oxfordhb/9780199930418.013.17. For an excellent short biography, see Gleick, J. (2004) *Isaac Newton*. New York, NY: Vintage.
4   Armbruster, S.W. (2001) Evolution of floral form: electrostatic forces, pollination, and adaptive compromise. *New Phytologist* 152: 181–3. https://doi.org/10.1046/j.0028-646X.2001.00268.x
5   Iversen, P. and Lacks, D.J. (2012) A life of its own: the tenuous connection between Thales of Miletus and the study of electrostatic charging. *Journal of Electrostatics* 70 (3): 309–11. https://doi.org/10.1016/j.elstat.2012.03.002
6   The bipolar nature of charge was first detailed in Franklin, B. (1751) *Experiments and Observations on Electricity, Made at Philadelphia in America, by Mr. Benjamin Franklin, and Communicated in several Letters to Mr. P. Collinson, of London, F.R.S.* London: E. Cave. Franklin's comments about the Bald Eagle were made in a letter to his daughter Sarah Bache, dated January 26, 1784.
7   Vaknin, Y., Gan-Mor, S., Bechar, A. *et al.* (2000) The role of electrostatic forces in pollination. *Plant Systematics and Evolution* 222: 133–42. https://doi.org/10.1007/BF00984099
8   Badger, M., Ortega-Jimenez, V.M., von Rabenau, L. *et al.* (2015) Electrostatic charge on flying hummingbirds and its potential role in pollination. *PLoS ONE* 10 (9): e0138003. https://doi.org/10.1371/journal.pone.0138003
9   Lowrie, W. (2007) *Fundamentals of Geophysics*. Cambridge: Cambridge University Press.
10 Morford, J., Jaggers, P., Guilford, T. *et al.* (2022) Magnetic stop signs signal a European songbird's arrival at the breeding site after migration. *Science* 375 (6579): 446–9. https://doi.org/10.1126/science.abj4210
11 Pearle, P., Collett, B., Bart, K. *et al.* (2010) What Brown saw and you can too. *American Journal of Physics* 78: 1278–89. https://doi.org/10.1119/1.3475685
12 Chowdhury, D. (2005) 100 years of Einstein's theory of Brownian motion: from pollen grains to protein trains. *arXiv:cond-mat*. https://doi.org/10.48550/arXiv.cond-mat/0504610
13 From Volume I, Chapter 41 of Feynman, R.P., Leighton, R.B. and Sands, M. (2011) *The Feynman Lectures on Physics*. New Millennium edition. New York, NY: Basic Books.
14 Wulf, A. (2016) *The Invention of Nature: Alexander von Humboldt's New World*. New York, NY: Vintage.

Chapter 3 On the Open Seas: Length Scales, Migration and Molecules

1   Ehrlich, P., Dobkin, D.S. and Wheye, D. (1988) *The Birder's Handbook: A Field Guide to the Natural History of North American Birds*. New York, NY: Simon and Schuster.
2   Fijn, R.C., Hiemstra, D., Phillips, R.A. *et al.* (2013) Arctic Terns *Sterna paradisaea* from the Netherlands migrate record distances across three oceans to Wilkes Land, East Antarctica. *Ardea* 101 (1): 3–12. https://doi.org/10.5253/078.101.0102
3   See a tracker such as: https://www.usdebtclock.org or https://fiscaldata.treasury.gov/americas-finance-guide/national-debt/
4   An online search for 'google earth alternatives' will reveal a variety of similar applications.
5   The mouth of the Ganges, at the north of the Bay of Bengal, is a lovely place for a virtual visit from above. Here, the great river splits outward, breaking up into ever smaller

branching and bending arms, smooth and flowing. In stark contrast are the tributaries flowing into the Nile in southern Egypt, with jagged, lightning-like zigzags. The coasts of Greenland and Norway are obvious choices for ragged rock structures cut up in big, medium and small grooves by flowing ice or water. Check out the lakes in Nunavut, Canada; and the vermiculated ridges and rows of sand dunes in Western Mauritania.

6   Sigee, D.C. (2005) *Freshwater Microbiology*. West Sussex: John Wiley and Sons.
7   To see just how beautifully we can now image individual molecules, see some recent work on using atomic force microscopy, such as Carracedo-Cosme, J. and Pérez, R. (2024) Molecular identification with atomic force microscopy and conditional generative adversarial networks. *NPJ Computational Materials* 10: 19. https://doi.org/10.1038/s41524-023-01179-1
8   Zhu, F., Xu, P. and Zong, J. (2023) Moore's Law: the potential, limits, and breakthroughs. *Applied and Computational Engineering* 10: 307–15. https://doi.org/10.54254/2755-2721/10/20230038
9   From one of the best introductions to physics ever written. If you only ever read one chapter of *The Feynman Lectures* (available for free online) it should be Volume 1, Chapter 1. Feynman, R.P., Leighton, R.B. and Sands, M. (2011) *The Feynman Lectures on Physics*, New Millennium edition. New York, NY: Basic Books.
10 Alberts, B., Bray, D., Hopkin, K. *et al.* (2014) *Essential Cell Biology*, 4th Edition. New York, NY: Garland Science.
11 Fánge, R., Schmidt-Nielsen, K. and Osaki, H. (1958) The salt gland of the Herring Gull. *The Biological Bulletin* 115: 162–71.
12 Braun, E.J. (2015) Chapter 12—Osmoregulatory Systems of Birds. In Scanes, C.G. (ed.) *Sturkie's Avian Physiology*, 6th edition. London: Academic Press.
13 The fascinating business of how ATP powers cells is described in any biochemistry text, such as Voet, D. and Voet, J.G. (2011) *Biochemistry*, 4th Edition. New York, NY: Wiley.
14 Pearson, K. (1905) The problem of the random walk. *Nature* 72: 294. https://doi.org/10.1038/072294b0
15 Fama, E.F. (1965) Random walks in stock market prices. *Financial Analysts Journal* 21: 55–59; Codling, E.A., Plank, M.J. and Benhamou, S. (2008) Random walk models in biology. *Journal of the Royal Society Interface*, 5: 813–34. https://doi.org/10.1098/rsif.2008.0014; Besenczi, R., Bátfai, N., Jeszenszky, P. *et al.* (2021) Large-scale simulation of traffic flow using Markov model. *PLoS ONE* 16: e0246062. https://doi.org/10.1371/journal.pone.0246062
16 Fischer, J., Walter, W.D. and Avery, M. (2013) Brownian bridge movement models to characterize birds' home ranges. *The Condor* 115 (2): 298–305. https://doi.org/10.1525/cond.2013.110168; Palm, E.C., Newman, S.H., Prosser, D.J. *et al.* (2015) Mapping migratory flyways in Asia using dynamic Brownian bridge movement models. *Movement Ecology* 3 (1): 3. https://doi.org/10.1186/s40462-015-0029-6
17 For a highly readable introduction and summary of Lévy flights and their use by seabirds and other animals, see Reynolds, A.M. (2018) Current status and future directions of Lévy walk research. *Biology Open* 7: bio030106. https://doi.org/10.1242/bio.030106
18 Humphries, N. and Sims, D. (2014) Optimal foraging strategies: Lévy walks balance searching and patch exploitation under a very broad range of conditions. *Journal of Theoretical Biology* 358: 179–93. https://doi.org/10.1016/j.jtbi.2014.05.032
19 Viswanathan, G.M., Afanasyev, V., Buldyrev, S.V. *et al.* (1996) Lévy flight search patterns of wandering albatrosses. *Nature* 381: 413–15. https://doi.org/10.1038/381413a0

20 Edwards, A., Phillips, R., Watkins, N. *et al.* (2007) Revisiting Lévy flight search patterns of wandering albatrosses, bumblebees and deer. *Nature* 449: 1044–48. https://doi.org/10.1038/nature06199
21 Humphries, N.E., Weimerskirch, H., Queiroz, N. *et al.* (2012) Foraging success of biological Lévy flights recorded in situ. *Proceedings of the National Academy of Sciences of the United States of America* 109 (19): 7169–74. https://doi.org/10.1073/pnas.1121201109; Focardi, S. and Cecere, J.G. (2014) The Lévy flight foraging hypothesis in a pelagic seabird. *The Journal of Animal Ecology* 83: 353–64. https://doi.org/10.1111/1365-2656.12147; Sims, D., Southall, E., Humphries, N. *et al.* (2008) Scaling laws of marine predator search behaviour. *Nature* 451: 1098–1102. https://doi.org/10.1038/nature06518
22 Reynolds, A., Bartumeus, F., Kölzsch, A. *et al.* (2016) Signatures of chaos in animal search patterns. *Scientific Reports* 6: 23492. https://doi.org/10.1038/srep23492
23 Colorni, A., Dorigo, M. and Maniezzo, V. (1991) Distributed optimization by ant colonies. *Proceedings of the First European Conference on Artificial Life* 142: 134–42.
24 For a fairly scathing review, see Sörensen, K. (2013) Metaheuristics—the metaphor exposed. *International Transactions of Operations Research* 22: 3–18. https://doi.org/10.1111/itor.12001. A more forgiving take is given by Lones, M.A. (2020) Mitigating metaphors: a comprehensible guide to recent nature-inspired algorithms, *SN Computer Science* 1: 49. https://doi.org/10.1007/s42979-019-0050-8
25 Liu, T. and Yu, Z. (2022) The analysis of financial market risk based on machine learning and particle swarm optimization algorithm. *EURASIP Journal on Wireless Communications and Networking* 31: 1–17. https://doi.org/10.1186/s13638-022-02117-3; Dhiman, G. and Chahar, V. (2018) Emperor Penguin optimizer: a bio-inspired algorithm for engineering problems. *Knowledge-Based Systems* 159: 20–50. https://doi.org/10.1016/j.knosys.2018.06.001; Yang, X.S. and Deb, S. (2009) Cuckoo search via Lévy flights. *Proceedings of World Congress on Nature and Biologically Inspired Computing (NaBIC 2009)*: 210–14. https://doi.org/10.1109/NABIC.2009.5393690; Yang, X.S. and Deb, S. (2010) Eagle strategy using Levy walk and firefly algorithms for stochastic optimization. In González, J.R., Pelta, D.A., Cruz, C. *et al.* (eds) *Nature Inspired Cooperative Strategies for Optimization (NICSO 2010). Studies in Computational Intelligence* 284: 101–11. Berlin: Springer. https://doi.org/10.1007/978-3-642-12538-6_9; Mohammadi-Balani, A., Nayeri, M.D., Azar, A. *et al.* (2021) Golden eagle optimizer: a nature-inspired metaheuristic algorithm. *Computers and Industrial Engineering* 152: 107050. https://doi.org/10.1016/j.cie.2020.107050; Heidari, A.A., Mirjalili, S., Faris, H. *et al.* (2019) Harris hawks optimization: Algorithm and applications. *Future Generation Computer Systems* 97: 849–72. https://doi.org/10.1016/j.future.2019.02.028

Chapter 4 In a City Park: Movement, Murmuration and Magnetism

1 Marche S. (2012) *How Shakespeare Changed Everything*. New York, NY: Harper Perennial.
2 Fugate, L. and Miller, J.M. (2021) Shakespeare's starlings: literary history and the fictions of invasiveness. *Environmental Humanities* 13 (2): 301–22. https://doi.org/10.1215/22011919-9320167
3 West, M.J. and King, A.P. (1990) Mozart's starling. *American Scientist* 78: 106–14.
4 Hopkins wrote *Pied Beauty* in 1877, one of his many hymns to nature. Another line reads 'Whatever is fickle, freckled (who knows how?)'; Alan Turing would answer 'how' in 1952.

5 Turing, A. (1952) The chemical basis of morphogenesis. *Philosophical Transactions of the Royal Society of London B* 237: 37–72.
6 The reaction term used here was of the form $R_a = a(1-a) - ab^2$ and $R_b = ab^2 - (a+b)b$. Altering the values of a and b leads to different patterns.
7 Inaba, M. and Chuong, C.M. (2020) Avian pigment pattern formation: developmental control of macro- (across the body) and micro- (within a feather) level of pigment patterns. *Frontiers in Cell and Developmental Biology* 8: 620. https://doi.org/10.3389/fcell.2020.00620; Prum, R.O. and Williamson, S. (2002) Reaction–diffusion models of within-feather pigmentation patterning. *Proceedings of the Royal Society B* 269 (1493): 781–92. https://doi.org/10.1098/rspb.2001.1896
8 Harris, M.P., Williamson, S., Fallon, J.F. *et al.* (2005) Molecular evidence for an activator-inhibitor mechanism in development of embryonic feather branching. *Proceedings of the National Academy of Sciences of the United States of America* 102 (33): 11734–9. https://doi.org/10.1073/pnas.0500781102
9 Reynolds, C. (1987) Flocks, herds, and schools: a distributed behavioral model. *ACM SIGGRAPH Computer Graphics* 21 (4): 25–34. https://doi.org/10.1145/37402.37406
10 Carlson, W.E. (2017) *Computer Graphics and Computer Animation: A Retrospective Overview*. Columbus, OH: The Ohio State University. https://books.google.com/books?id=kpnxswEACAAJ
11 Anderson, P.W. (1972) More is different: broken symmetry and the nature of the hierarchical structure of science. *Science* 177 (4047): 393–6. https://doi.org/10.1126/science.177.4047.393
12 Johnson, G. (1999) *Strange Beauty: Murray Gell-Mann and the Revolution in Twentieth-Century Physics*. New York, NY: Knopf; Gell-Mann, M. (1994) *The Quark and the Jaguar: Adventures in the Simple and the Complex*. New York, NY: W.H. Freeman and Co.
13 Griffiths, D. (1987) *Introduction to Elementary Particles*, 2nd edition. New York, NY: John Wiley & Sons.
14 Gell-Mann wasn't the only one to find the solution. Another physicist, George Zweig, had the same idea independently, and called the composite particles 'aces'. Gell-Mann's name stuck. That Zweig did not share in the Nobel honors is widely considered a snub.
15 Johnson (1999).
16 Hobson, A. (2013) There are no particles, there are only fields. *American Journal of Physics* 8: 211–23. https://doi.org/10.1119/1.4789885
17 Sailer, T., Debierre, V., Harman, Z. *et al.* (2022) Measurement of the bound-electron g-factor difference in coupled ions. *Nature* 606: 479–83. https://doi.org/10.1038/s41586-022-04807-w
18 A simple, if imperfect, visualization involves fiddling with some coins—say, pennies—arranged in a circle with their edges touching. Five can form a small circle, but if we want a larger one, we will need to add a sixth. In this manner, we can construct a set of 'orbitals' that each have a different size. Note that the circles we can make have particular sizes, with no intermediates. In this analogy, the diameter of the coin corresponds to a wavelength. Of course, atoms are three-dimensional, and the details are more complex, but the basic concept is the same. Since we have the coins out, let's demonstrate the differences in mass between various parts of an atom. If an electron weighed as much as a penny, a single proton (or neutron) would come in at about 26 pounds, comparable to the weight of a two-year-old child. A carbon atom (with 12 nucleons) would weigh *over 300 pounds* and be comparable to a refrigerator.

19 The proper name for the technique that MRI uses is nuclear magnetic resonance (NMR). Apparently, the term 'nuclear' retains negative connotations for some. For an introduction to the technique, see Sprawls, P. (2001) *Magnetic Resonance Imaging: Principles, Methods, and Techniques*. Madison, WI: Medical Physics Publishing.
20 The idea that avian magnetoreception might exploit the radical pair mechanism was first proposed in Schulten, K., Swenberg, C.E. and Weller, A. (1978) A biomagnetic sensory mechanism based on magnetic field modulated coherent electron spin motion. *Zeitschrift für Physikalische Chemie* 111: 1–5. https://doi.org/10.1524/zpch.1978.111.1.001. For a full overview of the chemistry see Hore, P.J. and Mouritsen, H. (2016) The radical-pair mechanism of magnetoreception. *Annual Review of Biophysics* 45 (1): 299–344. https://doi.org/10.1146/annurev-biophys-032116-094545
21 Nießner, C., Gross, J.C., Denzau, S. *et al.* (2016) Seasonally changing cryptochrome 1b expression in the retinal ganglion cells of a migrating passerine bird. *PLoS ONE* 11: e0150377. https://doi.org/10.1371/journal.pone.0150377
22 Stapput, K., Thalau, P., Wiltschko, R. *et al.* (2008) Orientation of birds in total darkness. *Current Biology* 18 (8): 602–06. https://doi.org/10.1016/j.cub.2008.03.046
23 Hiscock, H.G., Mouritsen, H., Manolopoulos, D.E. *et al.* (2017) Disruption of magnetic compass orientation in migratory birds by radiofrequency electromagnetic fields. *Biophysical Journal* 113 (7): 1475–84. https://doi.org/10.1016/j.bpj.2017.07.031
24 Muheim R., Sjöberg S. and Pinzon-Rodriguez A. (2016) Polarized light modulates light-dependent magnetic compass orientation in birds. *Proceedings of the National Academy of Sciences* 113 (6): 1654–9. https://doi.org/10.1073/pnas.1513391113

Chapter 5 By a Forest Pond: Impacts, Waves and Sounds

1 This equation is not universal. Drop a marble into a thick gel, and the force will depend on velocity, not its square. For the motions that concern us, this expression is appropriate.
2 Ropert-Coudert, Y., Grémillet, D., Ryan, P. *et al.* (2004) Between air and water: the plunge dive of the Cape Gannet *Morus capensis. Ibis* 146 (2): 281–290. https://doi.org/10.1111/j.1474-919x.2003.00250.x
3 Crandell, K., Howe, R. and Falkingham, P.L. (2019) Repeated evolution of drag reduction at the air–water interface in diving kingfishers. *Journal of The Royal Society Interface* 16: 20190125. https://doi.org/10.1098/rsif.2019.0125
4 Sharker, S., Holekamp, S., Mansoor, M. *et al.* (2019) Water entry impact dynamics of diving birds. *Bioinspiration and Biomimetics* 14 (5): 056013. https://doi.org/10.1088/1748-3190/ab38cc
5 May, P.R.A., Fuster, J., Haber, J. *et al.* (1979) Woodpecker drilling behavior. An endorsement of the rotational theory of impact brain injury. *Archives of Neurology* 36 (3): 370–3. https://doi.org/10.1001/archneur.1979.00500420080011
6 Van Wassenbergh, S. and Mielke, M. (2024) Why woodpeckers don't get concussions. *Physics Today* 77 (1): 54–5. https://doi.org/10.1063/PT.3.5385
7 Biewener, A.A. (2022) Physiology: woodpecker skulls are not shock absorbers. *Current Biology* 32 (14): 767–9. https://doi.org/10.1016/j.cub.2022.06.037
8 Van Wassenbergh, S., Ortlieb, E.J., Mielke, M. *et al.* (2022) Woodpeckers minimize cranial absorption of shocks. *Current Biology* 32 (14): 3189–94. https://doi.org/10.1016/j.cub.2022.05.052

9 Gibson, L.J. (2006) Woodpecker pecking: how woodpeckers avoid brain injury. *Journal of Zoology* 270 (3): 462–5. https://doi.org/10.1111/j.1469-7998.2006.00166.x

10 Vincent, L., Xiao, T., Yohann, D. *et al.* (2017) The dynamics of water entry. *Journal of Fluid Mechanics* 846: 508–35. https://doi.org/10.1017/jfm.2018.273

11 The profile of water waves is almost never a single, perfect sine wave. The details depend on many factors, including the depth of the water, the time elapsed, and so on.

12 The lowest intensity we can hear, I_0 is about 10–12 W/m². To express any other intensity I on the dB scale, we use the common logarithm of its ratio to I_0: dB = $10\log(I/I_0)$. On this scale, 10 dB means 10 times louder, 20 dB means 100 times louder, 30 dB means 1,000 times louder, and so on.

13 The discussion here is drawn from the excellent and thorough treatment by Mindlin, G.B. and Laje, R. (2005) *The Physics of Birdsong (Biological and Medical Physics, Biomedical Engineering)*. New York, NY: Springer.

14 Riede, T. and Goller, F. (2010) Functional morphology of the sound-generating labia in the syrinx of two songbird species. *Journal of Anatomy* 216: 23–36. https://doi.org/10.1111/j.1469-7580.2009.01161.x

15 We will treat this in more detail in Chapter 12, in the context of flight.

16 Fletcher, N.H. and Rossing, T.D. (1991) *The Physics of Musical Instruments*. New York, NY: Springer-Verlag.

17 Boutin, H., Smith, J. and Wolfe, J. (2020) Trombone lip mechanics with inertive and compliant loads ('lipping up and down'). *Journal of the Acoustical Society of America* 147 (6): 4133–44. https://doi.org/10.1121/10.0001466

18 Zolliger, S.A., Riede, T. and Suthers, R.A. (2008) Two-voice complexity from a single side of the syrinx in northern mockingbird *Mimus polyglottos* vocalizations. *Journal of Experimental Biology* 211 (12): 1978–91. https://doi.org/10.1242/jeb.014092

19 Nowicki, S. and Capranica, R.R. (1986) Bilateral syringeal coupling during phonation of a songbird. *The Journal of Neuroscience* 6 (12): 3595–610. https://doi.org/10.1523/jneurosci.06-12-03595.1986

20 Fletcher, N., Riede, T., Beckers, G. *et al.* (2005) Vocal tract filtering and the 'coo' of doves. *The Journal of the Acoustical Society of America* 116 (6): 3750–6. https://doi.org/10.1121/1.1811491

21 Brinkløv, S., Fenton, M.B. and Ratcliffe, J.M. (2013) Echolocation in Oilbirds and swiftlets. *Frontiers in Physiology* 4: 123. https://doi.org/10.3389/fphys.2013.00123

Chapter 6 Under Night's Cover: Hearing, Recording and Analyzing Birdsong

1 Singling out just a few species to praise for their otherworldly sounds. In reality, there is something magnificent in every bird song.

2 Normal human hearing goes from 20 Hz to 20,000 Hz. The frequencies of visible light, which we will discuss in detail in Chapter 7, range from about 400 to 800 THz (1 THz = 10^{12} Hz).

3 The discussion of a bird's auditory system here references the extensive treatment of Köppl, C. (2015) Chapter 6—Avian Hearing. In Scanes, C.G. (ed') *Sturkie's Avian Physiology*, 6th edition. London: Academic Press.

4   Hamra, M., Shinnawi, S., Vaizer, M.C. et al. (2020) Rapid imaging of tympanic membrane vibrations in humans. *Biomedical Optics Express* 11 (11): 6470–9. https://doi.org/10.1364/BOE.402097
5   Gan, R.Z., Reeves, B.P. and Wang, X. (2007) Modeling of sound transmission from ear canal to cochlea. *Annals of Biomedical Engineering* 35 (12): 2180–95. https://doi.org/10.1007/s10439-007-9366-y
6   Rubel, E.W., Furrer, S.A. and Stone, J.S. (2013) A brief history of hair cell regeneration research and speculations on the future. *Hearing Research* 297: 42–51. https://doi.org/10.1016/j.heares.2012.12.014
7   Köppl (2015).
8   Coles, R.B. and Guppy, A. (1987) Directional hearing in the barn owl (*Tyto alba*). *Journal of Comparative Physiology A* 163 (1): 117–33. https://doi.org/10.1007/BF00612002
9   von Campenhausen, M. and Wagner, H. (2006) Influence of the facial ruff on the sound-receiving characteristics of the Barn Owl's ears. *Journal of Comparative Physiology A* 192 (10): 1073–82. https://doi.org/10.1007/s00359-006-0139-0
10 Thomas Edison's phonograph patent in 1877 provides a convenient place to mark the beginning of recording and playback. See Read, O. and Welch W.L. (1976) *From Tin Foil to Stereo: Evolution of the Phonograph*. Indianapolis, IN: Howard W. Sams. For a review of recording technology specific to bioacoustics, see Erbe, C. and Thomas, J.A. (eds) (2022) *Exploring Animal Behavior Through Sound: Volume 1 Methods*. Cham: Springer Nature. https://doi.org/10.1007/978-3-030-97540-1
11 Although, as previously noted, Mozart transcribed his starling's song onto the staff, with the annotation, 'That was beautiful!'
12 A photo of this rig appears in Erbe and Thomas (2022).
13 Consulting a Radio Shack catalog from 1967, one will find that a commercially available tape recorder weighing four pounds and powered by four 'C' batteries provided 60 minutes of recording per cassette and cost $89.50. This is estimated to equal about $800 in 2024 dollars.
14 Rayburn, R.A. (2012) *Eargle's The Microphone Book*, 3rd edition. Burlington, MA: Focal Press.
15 Zawawi, S.A., Hamzah, A.A., Majlis, B.Y. et al. (2020) A review of MEMS capacitive microphones. *Micromachines* 11 (5): 484–510. https://doi.org/10.3390/mi11050484
16 A classic paper on the use of a dish specifically for capturing bird song is: Wahlström, S. (1985) The parabolic reflector as an acoustical amplifier. *Journal of The Audio Engineering Society* 33 (6): 418–29. http://www.aes.org/e-lib/browse.cfm?elib=4443
17 González-García, F., Sosa-López, J.R., Ornelas, J.F. et al. (2016) Individual variation in the booming calls of captive Horned Guans (*Oreophasis derbianus*): an endangered Neotropical mountain bird. *Bioacoustics* 26 (2): 185–98. https://doi.org/10.1080/09524622.2016.1233513
18 Diffraction is a result of interference, and might be stated simply as the ability of waves to 'bend around corners'. When waves pass through an opening, every point in that opening acts as a source of waves. For long wavelengths coming out of a small aperture, the different distances that the various waves travel as they spread out are much smaller than the wavelength. This means that the waves will spread out everywhere and stay more-or-less in phase and enjoy constructive interference. When short wavelengths come out of a large aperture, they will maintain the same phase with one another, only as they travel straight ahead; to the sides there will be much destructive interference, because different

distances they travel result in very different phases. It is very similar to the situation shown in Figure 5.3.
19 Bai, M. and Lo, Y. (2013) Refined acoustic modeling and analysis of shotgun microphones. *The Journal of the Acoustical Society of America* 133 (4): 2036–45. https://doi.org/10.1121/1.4792147
20 Smith, S.W. (1999) *The Scientist and Engineer's Guide to Digital Signal Processing*. San Diego, CA: California Technical Publishing.
21 Tiwari, S. (2016) An introduction to QR code technology. *2016 International Conference on Information Technology (ICIT)*, Bhubaneswar, India 39–44. https://doi.org/10.1109/ICIT.2016.021
22 Langton, C., Levin, V., Tishman, R. *et al.* (2017) *The Intuitive Guide to Fourier Analysis & Spectral Estimation*. Chicago, IL: Mountcastle Academic.
23 Heideman, M.T., Johnson, D.H. and Burrus, C.S. (1984) Gauss and the history of the fast Fourier transform. *IEEE ASSP Magazine* 1 (4): 14–21. https://doi.org/10.1109/MASSP.1984.1162257
24 Kahl, S., Wood, C.M., Eibl, M. *et al.* (2021) BirdNET: A deep learning solution for avian diversity monitoring. *Ecological Informatics* 61: 101236. https://doi.org/10.1016/j.ecoinf.2021.101236
25 The leading repository is doubtlessly https://www.xeno-canto.org. This free resource curates not only bird calls and songs, but the sounds of all animals.

Chapter 7 At the Lake: Sunlight, Reflection and Refraction

1 We attempted in Chapter 1 to develop a sense for large numbers, but it is very difficult to think up a way to intuitively grapple with 15 million degrees.
2 Deuterons don't readily form by proton–neutron collisions because there are very few free neutrons running around; they are unstable when left alone and decay into protons.
3 There is an ongoing debate—a truly academic one—about the possibility of life in a 'weakless' universe. Only by changing other features of nature would this seem plausible, as argued in Grohs, E., Howe, A. and Adams, F. (2018) Universes without the weak force: astrophysical processes with stable neutrons. *Physical Review D* 97 (4): 043003. https://doi.org/10.1103/PhysRevD.97.043003
4 Fraknoi, A., Morrison, D. and Wolff, S. (2022) *Astronomy*, 2nd edition. Houston, TX: OpenStax.
5 Johnson, N.K. and Cicero, C. (2004) New mitochondrial DNA data affirm the importance of Pleistocene speciation in North American birds. *Evolution* 58 (5): 1122–30. https://doi.org/10.1111/j.0014-3820.2004.tb00445.x
6 The term 'spectrum' long predates its application to sound; it was first used by Isaac Newton to describe the range of light colors that he famously studied using a prism.
7 Data made available by the National Renewable Energy Laboratory at https://www.nrel.gov/grid/solar-resource/spectra-am1.5.html
8 Horváth, G. (ed.) (2014) *Polarized Light and Polarization Vision in Animal Sciences*. New York, NY: Springer.
9 Muheim, R., Sjöberg, S. and Pinzon-Rodriguez, A. (2016) Polarized light modulates light-dependent magnetic compass orientation in birds. *Proceedings of the National*

Academy of Sciences of the United States of America 113 (6): 1654–9. https://doi.org/10.1073/pnas.1513391113

10 Muheim, R., Phillips, J. and Akesson, S. (2006) Polarized light cues underlie compass calibration in migratory songbirds. *Science* 313: 837–9. https://doi.org/10.1126/science.1129709

11 Also known as Fermat's principle, as it was first formulated by French mathematician Pierre de Fermat. It is an early example of a variational principle, in which one looks for variations on a path in order to select the route that minimizes some quantity: in this case, transit time.

12 For a full derivation of this effect, see Volume I, Chapter 31 of Feynman, R.P., Leighton, R.B. and Sands, M. (2011) *The Feynman Lectures on Physics*, New Millennium edition. New York, NY: Basic Books. https://www.feynmanlectures.caltech.edu/I_31.html

13 Hareli, S., Schechtman, E. and Arad, Z. (1999) Cattle egrets are less able to cope with light refraction than are other herons. *Animal Behaviour* 57 (3): 687–94. https://doi.org/10.1006/anbe.1998.1002

14 Katzir, G. and Intrator, N. (1987) Striking of underwater prey by a reef heron, *Egretta gularis schistacea*. *Journal of Comparative Physiology A* 160: 517–23. https://doi.org/10.1007/BF00615085

15 Brown, H.K.M., Rubega, M. and Dierssen, H.M. (2021) The light's in my eyes: optical modeling demonstrates wind is more important than sea surface-reflected sunlight for foraging herons. *PeerJ* 9: e12006. https://doi.org/10.7717/peerj.12006

16 Rajchard, J. (2018) Ultraviolet (UV) light perception by birds: a review. *Veterinarni Medicina* 54 (8): 351–9. https://doi.org/10.17221/110/2009-VETMED

17 Eaton, M.D. (2007) Avian visual perspective on plumage coloration confirms rarity of sexually monochromatic North American passerines. *The Auk* 124 (1): 155–61. https://doi.org/10.1093/auk/124.1.155

18 The benefits of UV may extend beyond plumage signaling, as shown in Tedore, C. and Nilsson, D. (2019) Avian UV vision enhances leaf surface contrasts in forest environments. *Nature Communications* 10: 238. https://doi.org/10.1038/s41467-018-08142-5. A recent study with a variety of images is Tunes, P., Camargo, M.G. and Guimarães, E. (2021) Floral UV features of plant species from a Neotropical savanna. *Frontiers in Plant Science* 12: 618028. https://doi.org/10.3389/fpls.2021.618028. For a comparison of effects on insects and birds, see Papiorek, S., Junker, R., Alves-dos-Santos, I. *et al.* (2015) Bees, birds and yellow flowers: pollinator-dependent convergent evolution of UV-patterns. *Plant Biology* 18 (1): 46–55. https://doi.org/10.1111/plb.12322

19 Zampiga, E., Gaibani, G., Csermely, D. *et al.* (2006) Innate and learned aspects of vole urine UV-reflectance use in the hunting behaviour of the common kestrel *Falco tinnunculus*. *Journal of Avian Biology* 37(4): 318–22. https://doi.org/10.1111/j.2006.0908-8857.03825.x

Chapter 8 During a Big Day: Light, Matter and Feather Colors

1 For a fantastic exploration of the myriad light effects that the inanimate world alone provides, see Lynch D.K. and Livingston W. (2001) *Color and Light in Nature*, 2nd edition. Cambridge: Cambridge University Press.

2 For a vivid demonstration of the effect, an online search for 'Fechner color' will provide plenty of simulations. For a general reference, see Krantz, J. (2020) Fechner's Colors and

Benham's Top. In Shamey, R. (ed.) *Encyclopedia of Color Science and Technology*. Berlin: Springer. https://doi.org/10.1007/978-3-642-27851-8_65-3
3   Foth, C. (2020) Introduction to the morphology, development, and ecology of feathers. In Foth, C. and Rauhut, O. (eds) *The Evolution of Feathers*. Cham: Springer Nature. https://doi.org/10.1007/978-3-030-27223-4_1
4   Parnell, A., Washington, A., Mykhaylyk, O. *et al.* (2015) Spatially modulated structural colour in bird feathers. *Scientific Reports* 5: 18317. https://doi.org/10.1038/srep18317. For a general account of feather structure, see Proctor, N.S. and Lynch, P.J. (1993) *Manual of Ornithology: Avian Structure and Function*, 1st edition. New Haven, CT: Yale University Press.
5   Dunning, J., Anvay, P., D'Alba, L. *et al.* (2023) How woodcocks produce the most brilliant white plumage patches among the birds. *Journal of the Royal Society Interface* 20 (200): 20220920. https://doi.org/10.1098/rsif.2022.0920
6   https://creativecommons.org/licenses/by/4.0/
7   An overview of this history is given by Noh, H., Liew, S.F., Saranathan, V. *et al.* (2010) How noniridescent colors are generated by quasi-ordered structures of bird feathers. *Advanced Materials* 22: 2871–80. https://doi.org/10.1002/adma.200903699
8   Prum, R.O. and Torres, R.H. (2003) A fourier tool for the analysis of coherent light scattering by bio-optical nanostructures. *Integrative and Comparative Biology* 43 (4): 591–602. https://doi.org/10.1093/icb/43.4.591
9   Hecht, E. (2017) *Optics*, 5th edition. Harlow: Pearson Education Limited.
10 Raman, C.V. (1934) The origin of the colours in the plumage of birds. In *Proceedings of the Indian Academy of Sciences-Section A* 1: 1–7. New Delhi: Springer India.
11 Proctor and Lynch (1993).
12 Prum, R.O., Torres, R.H., Williamson, S. *et al.* (1998) Coherent light scattering by blue feather barbs. *Nature* 396: 28–9. https://doi.org/10.1038/23838
13 Proctor and Lynch (1993).
14 Saranathan, V., Forster, J.D., Noh, H. *et al.* (2012) Structure and optical function of amorphous photonic nanostructures from avian feather barbs: a comparative small angle X-ray scattering (SAXS) analysis of 230 bird species. *Journal of the Royal Society Interface* 9: 2563–80. https://doi.org/10.1098/rsif.2012.0191
15 Greenewalt, C.H. (1991) *Hummingbirds*. Mineola, NY: Dover Publications.
16 Nordén, K.K., Faber, J.W., Babarović, F. *et al.* (2019) Melanosome diversity and convergence in the evolution of iridescent avian feathers-Implications for paleocolor reconstruction. *Evolution* 73: 15–27. https://doi.org/10.1111/evo.13641
17 Greenewalt (1991); Dyck J. (1987) Structure and light reflection of green feathers of fruit doves (*Ptilinopus* spp.) and an imperial pigeon (*Ducula concinna*). *Biologiske Skrifter* 30: 1–43; Prum, R.O. (1999) The anatomy and physics of avian structural colours. In Adams, N.J. and Slotow, R.H. (eds) *Proceedings of the 22nd International Ornithological Congress*: 1633–53. Durban: BirdLife South Africa; Harvey, T.A., Bostwick, K.S. and Marschner, S. (2013) Directional reflectance and milli-scale feather morphology of the African Emerald Cuckoo, *Chrysococcyx cupreus*. *Journal of the Royal Society Interface* 10: 20130391. https://doi.org/10.1098/rsif.2013.0391
18 Prum, R.O. (1999); Dyck, J. (1992) Reflectance spectra of plumage areas colored by green feather pigments. *The Auk* 109: 293–301.
19 Maia, L., Fleury, B., Lages, B. *et al.* (2011) Identification of reddish pigments in octocorals by Raman spectroscopy. *Journal of Raman Spectroscopy* 42: 653–8. https://doi.org/10.1002/jrs.2758

20 Bruice, P.Y. (2013) *Organic Chemistry*, 7th edition. London: Pearson.
21 Johnson, M.P. (2016) Photosynthesis. *Essays In Biochemistry* 60 (3): 255–73. https://doi.org/10.1042/EBC20160016
22 LaFountain, A.M., Prum, R.O. and Frank, H.A. (2015) Diversity, physiology, and evolution of avian plumage carotenoids and the role of carotenoid–protein interactions in plumage color appearance. *Archives of Biochemistry and Biophysics* 572: 201–12. https://doi.org/10.1016/j.abb.2015.01.016
23 Meredith, P. and Sarna, T. (2006) The physical and chemical properties of eumelanin. *Pigment Cell Research* 19: 572–94. https://doi.org/10.1111/j.1600-0749.2006.00345.x
24 Grieco, C., Kohl, F.R., Hanes, A.T. et al. (2020) Probing the heterogeneous structure of eumelanin using ultrafast vibrational fingerprinting. *Nature Communications* 11: 4569. https://doi.org/10.1038/s41467-020-18393-w
25 Costa, T., Feldhaus, M., Vilhena, F. et al. (2014) Preparation, characterization, cytotoxicity and antioxidant activity of DOPA melanin modified by amino acids: melanin-like oligomeric aggregates. *Journal of the Brazilian Chemical Society* 26: 273–81. https://doi.org/10.5935/0103-5053.20140277
26 Ahmad, N., Alqahtani, H., Al-Terary, S. et al. (2019) Control of optical absorption and fluorescence spectroscopies of natural melanin at different solution concentrations. *Optical and Quantum Electronics* 51: 1–13. https://doi.org/10.1007/s11082-019-1936-3

Chapter 9 In the Blind: Images, Eyes and Cameras

1 Most biochemistry texts will show its huge conjugated bond structures and dual peak absorption spectrum. Or see Johnson, M.P. (2016) Photosynthesis. *Essays In Biochemistry* 60: 255–73. https://doi.org/10.1042/EBC20160016
2 von Lintig, J., Kiser, P.D., Golczak, M. et al. (2010) The biochemical and structural basis for trans-to-cis isomerization of retinoids in the chemistry of vision. *Trends in Biochemical Sciences* 35 (7): 400–10. https://doi.org/10.1016/j.tibs.2010.01.005
3 Cephalopods such as squid famously got the design right, with the cells facing the 'right' way and no blind spot.
4 An overview of the fundamental process is given by Palczewski, K. (2011) Chemistry and Biology of Vision. *The Journal of Biological Chemistry*. 287 (3): 1612–9. https://doi.org/10.1074/jbc.R111.301150
5 Marieb, E.N. and Keller, S.M. (2021) *Essentials of Human Anatomy and Physiology*, 13th edition. Harlow: Pearson.
6 de Grip, W.J. and Ganapathy, S. (2022) Rhodopsins: An excitingly versatile protein species for research, development and creative engineering. *Frontiers in Chemistry* 10: 879609. https://doi.org/10.3389/fchem.2022.879609
7 Comparing the specified resolution and sensor dimensions allows for an easy estimate of the individual pixel size.
8 First used in the 5th century BCE, per Hammond, J.H. (1981) *The Camera Obscura: A Chronicle*. Boca Raton, FL: CRC Press.
9 A good general reference is Johnson, C., Jr. (2017) *Science for the Curious Photographer: An Introduction to the Science of Photography*, 2nd edition. New York, NY: Routledge.
10 Referring to the manufacturer's technical specifications is the only way to be certain about what the ISO settings mean.

11 As of 2022, some pixels are no larger than the wavelength of green light, giving an area of a quarter of a square micron. See Moore, S.K. (2022) Samsung and Omnivision claim smallest camera pixels: each has made CMOS camera chips with pixel pitches of just 0.56 micrometers. *IEEE Spectrum*, 10 Mar. 2022. https://spectrum.ieee.org/cmos-image-sensor-pixel
12 Wood, C.A. (1917) *The Fundus Oculi of Birds, Especially as Viewed by the Ophthalmoscope: A Study in Comparative Anatomy and Physiology*. Chicago, IL: Lakeside Press.
13 Potier, S. (2020) Visual adaptations in predatory and scavenging diurnal raptors. *Diversity* 12 (10): 400. https://doi.org/10.3390/d12100400
14 Mitkus, M., Potier, S., Martin, G. *et al.* (2018) Raptor vision. In *Oxford Research Encyclopedia of Neuroscience*. Oxford: Oxford University Press. https://doi.org/10.1093/acrefore/9780190264086.013.232
15 Jones, M.P., Pierce, K.E. and Ward, D. (2007) Avian vision: a review of form and function with special consideration to birds of prey. *Journal of Exotic Pet Medicine* 16 (2): 69–87. https://doi.org/10.1053/j.jepm.2007.03.012
16 Mitkus, Potier, Martin *et al.* (2018).
17 Land, M. (1999) The roles of head movements in the search and capture strategy of a tern (*Aves, Laridae*). *Journal of Computational Physiology A* 184: 265–72. https://doi.org/10.1007/s003590050324
18 Potier, S., Mitkus, M. and Kelber, A. (2020) Visual adaptations of diurnal and nocturnal raptors. *Seminars in Cell and Developmental Biology* 106: 116–26. https://doi.org/10.1016/j.semcdb.2020.05.004
19 For raptors, see Mitkus, Potier, Martin *et al.* (2018). For humans, see Wells-Gray, E., Choi, S., Bries, A. *et al.* (2016) Variation in rod and cone density from the fovea to the mid-periphery in healthy human retinas using adaptive optics scanning laser ophthalmoscopy. *Eye* 30 (8): 1135–43. https://doi.org/10.1038/eye.2016.107
20 Frey, K., Zimmerling, B., Scheibe, P. *et al.* (2017) Does the foveal shape influence the image formation in human eyes? *Advanced Optical Technologies* 6 (5): 403–10. https://doi.org/10.1515/aot-2017-0043
21 Potier, S., Mitkus, M., Lisney, T.J. *et al.* (2020) Inter-individual differences in foveal shape in a scavenging raptor, the black kite *Milvus migrans*. *Scientific Reports* 10 (1): 6133. https://doi.org/10.1038/s41598-020-63039-y
22 Quain, J., Schafer, E.A., Sharpey, W. *et al.* (1878) *Quain's Elements of Anatomy*, 8th edition. New York, NY: William Wood.
23 Zueva, L., Makarov, V., Zayas-Santiago, A. *et al.* (2014) Müller cell alignment in bird fovea: possible role in vision. *Journal of Neuroscience and Neuroengineering* 3: 85–91. https://doi.org/10.1166/jnsne.2014.1104
24 Mitkus, M., Olsson, P., Toomey, M.B. *et al.* (2017) Specialized photoreceptor composition in the raptor fovea. *Journal of Computation Neurology* 525 (9): 2152–63. https://doi.org/10.1002/cne.24190
25 Jones, Pierce and Ward (2007).
26 Interestingly, birds have eliminated the vascular layer altogether, opting instead to use a membrane that extends into the jelly-like filling (the vitreous humor) of the eye. This *pecten* structure releases nutrients into the thick surrounding liquid, which disperse and eventually reach the rods and cones. See Brach, V. (1977) The functional significance of the avian pecten: a review. *The Condor* 79: 321–7. https://doi.org/10.2307/1368009
27 Potier (2020).

28 Jones, Pierce and Ward (2007).
29 Zueva, Makarov, Zayas-Santiago *et al.* (2014).
30 Jones, Pierce and Ward (2007).
31 Reymond, L. (1985) Spatial visual acuity of the eagle *Aquila audax*: a behavioural, optical and anatomical investigation. *Vision Research* 25 (10): 1477–91. https://doi.org/10.1016/0042-6989(85)90226-3
32 Potier, Mitkus and Kelber (2020).
33 A specific reflective layer, the *tapeta lucidum*, is known to be present in a variety of nocturnal animals. Various accounts of nightjars (such as Cleere, N. (1998) *Nightjars: A Guide to the Nightjars, Nighthawks, and Their Relatives*. New Haven, CT: Yale University Press) assert that all nightjars and perhaps all Caprimulgiformes have tapeta. Other literature isn't clear. One paper claims that 'to date, only one bird species has been definitively found to harbor a tapetum: the Japanese Nightjar'. See Vee, S., Barclay, G. and Lents, N. (2022) The glow of the night: the tapetum lucidum as a co-adaptation for the inverted retina. *BioEssays* 44 (10): 2200003. https://doi.org/10.1002/bies.202200003. A much older reference states 'In birds, the tapetum lucidum is apparently absent although a choroid is present.' Prum, R.O. (1999) The anatomy and physics of avian structural colours. In Adams, N.J. and Slotow, R.H. (eds) *Proceedings of the 22nd International Ornithological Congress*, 1633–53. Johannesburg: BirdLife South Africa.
34 Hadden, P.W. and Jie, Z. (2023) An overview of the penguin visual system. *Vision* 7 (1): 6–30. https://doi.org/10.3390/vision7010006
35 https://creativecommons.org/licenses/by/4.0/
36 Urban, I., Uwurukundo, X., Stumpf, D. *et al.* (2020) Amphibious vision—Optical design model of the hooded merganser eye. *Vision Research* 175: 75–84. https://doi.org/10.1016/j.visres.2020.05.011
37 López-Alcón, D., Marín-Franch, I., Fernández-Sánchez, V. *et al.* (2016) Optical factors influencing the amplitude of accommodation. *Vision Research* 141: 16–22. https://doi.org/10.1016/j.visres.2016.09.003

Chapter 10 From a Great Distance: Lenses, Binoculars and Scopes

1 Hecht, E. (2017) *Optics*, 5th edition. Harlow: Pearson Education Limited.
2 Dunn, R. (2011) *The Telescope: A Short History*. London: Conway.
3 See Hurben, M. (2024) *Roof Prisms and Phase Coatings*. Manuscript in preparation. https://doi.org/10.17605/OSF.IO/S27H8
4 Sarangan, S. (2020) *Optical Thin Film Design*, 1st edition. Boca Raton, FL: CRC Press.
5 When the refractive indices are the same, this becomes zero (no reflection). It can also never be any larger than one.
6 Hecht (2017).
7 https://www.schott.com/en-us/products/optical-glass-p1000267/downloads
8 The idea dates back to 1932. See Baltag, O. (2015) History of automatic focusing reflected by patents. *Science Innovation* 3: 1–17. https://doi.org/10.11648/j.si.20150301.11
9 For a review of various approaches, see Rawat, D. and Jyoti, S. (2011) Review of motion estimation and video stabilization techniques for hand held mobile video. *Signal & Image Processing International Journal (SIPIJ)* 2: 159–68. https://doi.org/10.5121/sipij.2011.2213

Chapter 11 Under Extremes: Heat, Cold and Thermoregulation

1. Lewden, A., Enstipp, M.R., Picard, B. et al. (2017) High peripheral temperatures in king penguins while resting at sea: thermoregulation versus fat deposition. *Journal of Experimental Biology* 220 (17): 3084–94. https://doi.org/10.1242/jeb.158980
2. Rezende, E. and Bacigalupe, L. (2015) Thermoregulation in endotherms: physiological principles and ecological consequences. *Journal of Comparative Physiology B* 185 (7): 709–27. https://doi.org/10.1007/s00360-015-0909-5
3. Daniel, R.M., Danson, M.J., Eisenthal, R. et al. (2007) The effect of temperature on enzyme activity: new insights and their implications. *Extremophiles* 12 (1): 51–9. https://doi.org/10.1007/S00792-007-0089-7
4. Ancel, A., Gilbert, C., Poulin, N. et al. (2015) New insights into the huddling patterns of emperor penguins. *Animal Behaviour* 110: 91–8. https://doi.org/10.1016/j.anbehav.2015.09.019
5. West, G. (1965) Shivering and heat production in wild birds. *Physiological Zoology* 38 (2): 111–20. https://doi.org/10.1086/physzool.38.2.30152817
6. Another mechanism for some is to use the evaporation of urine. Cabello-Vergel, J., Soriano-Redondo, A., Villegas, A. et al. (2021) Urohidrosis as an overlooked cooling mechanism in long-legged birds. *Scientific Reports* 11: 20018. https://doi.org/10.1038/s41598-021-99296-8
7. Périard, J.D., Eijsvogels, T.M.H. and Daanen, H.A.M. (2021) Exercise under heat stress: thermoregulation, hydration, performance implications, and mitigation strategies. *Physiological Reviews* 101 (4): 1873–9. https://doi.org/10.1152/physrev.00038.2020
8. Tattersall, G.J., Andrade, D.V. and Abe, A.S. (2009) Heat exchange from the toucan bill reveals a controllable vascular thermal radiator. *Science* 325 (5939): 468–70. https://doi.org/10.1126/science.1175553
9. Greenberg, R., Danner, R., Olsen, B. et al. (2012) High summer temperature explains bill size variation in salt marsh sparrows. *Ecography* 35 (2): 146–52. https://doi.org/10.1111/j.1600-0587.2011.07002.x
10. Ryeland, J., Weston, M.A. and Symonds, M.R.E. (2017) Bill size mediates behavioural thermoregulation in birds. *Functional Ecology* 31 (4): 885–93. https://doi.org/10.1111/1365-2435.12814
11. Symonds, M.R. and Tattersall, G.J. (2010) Geographical variation in bill size across bird species provides evidence for Allen's rule. *The American Naturalist* 176 (2): 188–97. https://doi.org/10.1086/653666
12. Nudds, R.L. and Oswald, S.A. (2007) An interspecific test of Allen's rule: evolutionary implications for endothermic species. *Evolution* 61 (12): 2839–48. https://doi.org/10.1111/j.1558-5646.2007.00242.x
13. He, J., Tu, J., Yu, J. et al. (2023) A global assessment of Bergmann's rule in mammals and birds. *Global Change Biology* 29 (18): 5199–210. https://doi.org/10.1111/gcb.16860
14. McQueen, A., Klaassen, M., Tattersall, G.J. et al. (2022) Thermal adaptation best explains Bergmann's and Allen's rules across ecologically diverse shorebirds. *Nature Communications* 13: 1–12. https://doi.org/10.1038/s41467-022-32108-3
15. Knapik, J.J., Reynolds, K.L. and Castellani, J.W. (2020) Frostbite: pathophysiology, epidemiology, diagnosis, treatment, and prevention. *Journal of Special Operations Medicine* 20 (4): 123–35. https://doi.org/10.55460/PDX9-BG8G

16 Gagge, A.P. and Gonzalez, R.R. (2011) Mechanisms of heat exchange: biophysics and physiology. In Terjung, R. (ed.) *Comprehensive Physiology*. New York, NY: John Wiley & Sons. https://doi.org/10.1002/cphy.cp040104
17 Midtgård, U. (1989) Circulatory adaptations to cold in birds. In Bech, C. and Reinertsen, R.E. (eds) (1989) *Physiology of Cold Adaptation in Birds*. NATO ASI Series, 173. Boston, MA: Springer.
18 Perry, S. (2008) A first look at how fish gills work. *The Journal of Experimental Biology*, 211 (22): 3519–21. https://doi.org/10.1242/jeb.013599
19 Midtgård (1989).
20 Anyanwu, C., Abdelrahman, G., Hussein, B. et al. (2022) Heat analysis of a vacuum flask. *The Journal of Engineering and Exact Sciences* 8: 15174–01e. https://doi.org/10.18540/jcecvl8iss11pp15174-01e
21 The microstructure of the filaments for highly insulating feathers helps them cohere into an air-filling structure. See D'Alba, L., Carlsen, T., Ásgeirsson, Á. et al. (2017) Contributions of feather microstructure to eider down insulation properties. *Journal of Avian Biology* 48 (8): 1150–7. https://doi.org/10.1111/jav.01294
22 The 'preen oil' from the uropygial gland that many birds apply to their feathers has hydrophobic properties, but feathers can be quite waterproof without it. Moreover, not all birds have this gland; see Moyer, B.R., Rock, A.N. and Clayton, D.H. (2003) Experimental test of the importance of preen oil in Rock Doves (*Columba livia*). *The Auk* 120 (2): 490–6. https://doi.org/10.1093/auk/120.2.490. See also Montalti, D. and Salibian, A. (2000) Uropygial gland size and avian habitat. *Ornitologia Neotropical* 11 (4): 297–306. https://digitalcommons.usf.edu/ornitologia_neotropical/vol11/iss4/2
23 Cassie, A.B.D. and Baxter, S. (1944) Wettability of porous surfaces. *Transactions of The Faraday Society* 40: 546–51.
24 Rijke, A. and Jesser, W. (2011) The water penetration and repellency of feathers revisited. *The Condor* 113 (2): 245–54. https://doi.org/10.1525/cond.2011.100113
25 Rijke and Jesser (2011).
26 Terrill, R. and Shultz, A. (2022) Feather function and the evolution of birds. *Biological Reviews* 98 (2): 540–66. https://doi.org/10.1111/brv.12918
27 Zheng, K.D., Liang, D. and Wang, X. (2022) Contrasting coloured ventral wings are a visual collision avoidance signal in birds. *Proceedings of the Royal Society B* 289 (1978): 20220678. https://doi.org/10.1098/rspb.2022.0678
28 Galván, I., Rodríguez-Martínez, S. and Carrascal, L. (2018) Dark pigmentation limits thermal niche position in birds. *Functional Ecology* 32 (6): 1531–40. https://doi.org/10.1111/1365-2435.13094; Gunderson, A.R., Frame, A.M., Swaddle, J.P. et al. (2008) Resistance of melanized feathers to bacterial degradation: is it really so black and white? *Journal of Avian Biology* 39 (5): 539–45. https://doi.org/10.1111/j.0908-8857.2008.04413.x
29 Burtt, E.H., Jr. and Ichida, J. (2004) Gloger's rule, feather-degrading bacteria, and color variation among Song Sparrows. *The Condor* 106 (3): 681–6. https://doi.org/10.1093/condor/106.3.681. For a general summary, see Delhey, K. (2019) A review of Gloger's rule, an ecogeographical rule of colour: definitions, interpretations and evidence. *Biological Reviews* 94 (4): 1294–316. https://doi.org/10.1111/brv.12503
30 Bonser, R. (1995) Melanin and the abrasion resistance of feathers. *The Condor* 97 (2): 590–1. https://doi.org/10.2307/1369048

31 Dufour, P., Guerra Carande, J., Renaud, J. *et al.* (2020) Plumage colouration in gulls responds to their non-breeding climatic niche. *Global Ecology and Biogeography* 29 (10): 1704–15. https://doi.org/10.1111/geb.13142
32 Wolf, B. and Walsberg, G. (2000) The role of the plumage in heat transfer processes of birds. *American Zoologist* 40 (4): 575–84. https://doi.org/10.1093/icb/40.4.575
33 Ward, S., Rayner, J.M.V., Möller, U. *et al.* (1999) Heat transfer from starlings *Sturnus vulgaris* during flight. *Journal of Experimental Biology* 202 (12): 1589–602. https://doi.org/10.1242/jeb.202.12.1589
34 Léger, J. and Larochelle, J. (2006) On the importance of radiative heat exchange during nocturnal flight in birds. *The Journal of Experimental Biology* 209 (1): 103–14. https://doi.org/10.1242/jeb.01964
35 Weeks, B., Harvey, C., Tobias, J. *et al.* (2023) Bird wings are shaped by thermoregulatory demand for heat dissipation. *bioRxiv.* https://doi.org/10.1101/2023.02.06.527306
36 Ward, Rayner, Möller *et al.* (1999).
37 Léger and Larochelle (2006).
38 Rogalla, S., D'Alba, L., Verdoodt, A. *et al.* (2019) Hot wings: thermal impacts of wing coloration on surface temperature during bird flight. *Journal of the Royal Society Interface* 16 (156): 20190032. https://doi.org/10.1098/rsif.2019.0032

Chapter 12 Above the Earth: Wings, Lift and Flight

1 Carlson, J., Jaffe, A. and Wiles, A. (eds) (2006) *The Millennium Prize Problems.* Providence, RI: American Mathematical Society.
2 Wild, G. (2023) Is that lift diagram correct? A visual study of flight education literature. *Physics Education* 58 (3): 035018. https://doi.org/10.1088/1361-6552/acbad8
3 Soaring can occur on upward airflows that arise from other sources, notably when a wind strikes a vertical surface such as a cliff, the side of a ship, or even ocean waves; this effect is exploited by various seabirds.
4 This is the so-called 'sine-squared' rule of air resistance. See Liu, T. (2021) Evolutionary understanding of airfoil lift. *Advances in Aerodynamics* 3: 37. https://doi.org/10.1186/s42774-021-00089-4
5 A text written for pilots that argues that this effect is important for flight is Anderson, D.F. and Eberhardt, S. (2001) *Understanding Flight.* New York, NY: McGraw-Hill. You can demonstrate the effect by placing a cylinder such as a coffee can or large bottle between you and a lit candle. Blow onto the side of the bottle opposite to the candle; air that you set moving will not rebound as might be expected, but instead will cling to the sides of the object and follow its shape, until it arrives at the candle and extinguishes the flame.
6 Wild, G. (2023) Misunderstanding flight part 1: a century of flight and lift education literature. *Education Sciences* 13 (8): 762–98. https://doi.org/10.3390/educsci13080762
7 Babinsky, H. (2003) How do wings work? *Physics Education* 38 (6): 497–503. https://doi.org/10.1088/0031-9120/38/6/001
8 This is perhaps the most exasperating claim about lift that has gotten widespread traction. A good overview of this and similar problems is given by Wild (2023).
9 Babinsky (2003).
10 The clearest, most concise and most accurate treatment of flight that this author has encountered is McLean, D. (2018) Aerodynamic lift, part 1: the science; part 2: a

comprehensive physical explanation. *The Physics Teacher* 56 (8): 516–20; 521–4. https://doi.org/10.1119/1.5064558; https://doi.org/10.1119/1.5064559
11 As on the cover of Anderson and Eberhardt (2001).
12 Rogalla, S., D'Alba, L., Verdoodt, A. *et al.* (2019) Hot wings: thermal impacts of wing coloration on surface temperature during bird flight. *Journal of the Royal Society Interface* 16 (156): 20190032. https://doi.org/10.1098/rsif.2019.0032
13 Portugal, S., Hubel, T., Fritz, J. *et al.* (2014) Upwash exploitation and downwash avoidance by flap phasing in ibis formation flight. *Nature* 505: 399–402. https://doi.org/10.1038/nature12939
14 Similar in spirit to the drag coefficient we encountered before, it will depend on the airfoil shape, orientation, and other factors.
15 Tennekes, H. (2009) *The Simple Science of Flight*. Cambridge, MA: MIT Press, which includes plots of the 'Great Flight Diagram' ranging from insects through jumbo-jets.
16 Lee, S., Kim, J., Park, H. *et al.* (2015) The function of the alula in avian flight. *Scientific Reports* 5: 9914. https://doi.org/10.1038/srep09914
17 Something one can see many times without really noticing it or realizing the physical origin of the effect.
18 Also called lift-induced drag or vortex drag. One scenario in which this sometimes costly frictional effect is mitigated is when flying close (within a wingspan) to the ground. See, for example, Hainsworth, F.R. (1988) Induced drag savings from ground effect and formation flight in Brown Pelicans. *Journal of Experimental Biology* 135 (1): 431–44. https://doi.org/10.1242/jeb.135.1.431
19 Tennekes (2009); Denny, M. (2009) Dynamic soaring: aerodynamics for albatrosses. *European Journal of Physics* 30 (1): 75–84. https://doi.org/10.1088/0143-0807/30/1/008
20 Warrick, D., Hedrick, T., Fernández, M.J. *et al.* (2012) Hummingbird flight. *Current Biology* 22 (12): 472–7. https://doi.org/10.1016/j.cub.2012.04.057
21 Dvořák, R. (2016) Aerodynamics of bird flight. *EPJ Web of Conferences EFM15— Experimental Fluid Mechanics* 114: 01001. https://doi.org/10.1051/epjconf/201611401001
22 Dvořák (2016).
23 Warrick, Hedrick, Fernández *et al.* (2012).
24 Hedrick, T.L., Tobalske, B.W., Ros, I.G. *et al.* (2012) Morphological and kinematic basis of the hummingbird flight stroke: scaling of flight muscle transmission ratio. *Proceedings of the Royal Society B* 279: 1986–92. https://doi.org/10.1098/rspb.2011.2238
25 Pennycuick (1990); Pennycuick (1996); Pennycuick, C.J. (2001) Speeds and wingbeat frequencies of migrating birds compared with calculated benchmarks. *The Journal of Experimental Biology* 204 (19): 3283–94. https://doi.org/10.1242/jeb.204.19.3283
26 Pennycuick, C.J. (1990) Predicting wingbeat frequency and wavelength of birds. *The Journal of Experimental Biology* 150 (1): 171–85. https://doi.org/10.1242/jeb.150.1.171. It should be noted that a more refined (if less simple) model for wingbeat frequency later developed by Pennycuick can be expressed as

$$f = \frac{\beta m^{\frac{3}{8}}}{b^{\frac{23}{24}} S^{\frac{1}{3}}}$$

See Pennycuick, C.J. (1996) Wingbeat frequency of birds in steady cruising flight: new data and improved predictions. *The Journal of Experimental Biology* 199 (7): 1613–8. https://doi.org/10.1242/jeb.199.7.1613

Coda: Looking Back

1. Frank, M.G. and Conover, M.R. (2021) Diets of staging phalaropes at Great Salt Lake, Utah. *Wildlife Society Bulletin* 45: 27–35. https://doi.org/10.1002/wsb.1157
2. Mahoney, S.A. and Jehl, J.R. (1985) Adaptations of migratory shorebirds to highly saline and alkaline lakes: Wilson's Phalarope and American Avocet. *The Condor* 87(4): 520–7. https://doi.org/10.2307/1367950
3. Frank and Conover (2021).
4. Smith, W.G. (1889) The Wilson's Phalarope: *Phalaropus lobatus*. *The Ornithologists' and Oologists' Semi-Annual* 1(2): 14–5.
5. Obst, B.S., Hamner, W.M., Hamner, P.P. *et al.* (1996) Kinematics of phalarope spinning. *Nature* 384: 121.
6. Prakash, M., Quéré, D. and Bush, J. (2008) Surface tension transport of prey by feeding shorebirds: the capillary ratchet. *Science* 320: 931–4. https://doi.org/10.1126/science.1156023

Glossary

Abbe–König prism – a compound roof prism with four internal reflections, which is constructed from two cemented glass blocks. It is utilized in some binoculars to reorient the inverted, reversed image characteristic of a Keplerian telescope design without displacing the light off the optical axis. Unlike the related Schmidt–Pechan prism, it requires only anti-reflection and specialized phase coatings.

Absorption – the conversion of some or all light energy into the internal energy of the medium it enters.

Acceleration – the time rate of change of velocity. It is a vector quantity with both a magnitude and direction.

Accommodation – the reflexive process by which an eye adapts to form a focused image when the distance to an object changes, typically by modifying the shape, and hence the focal length, of the lens. Some diving birds may change the shape of their corneas to achieve the accommodation necessary for underwater vision.

Adenosine triphosphate (ATP) – an organic molecule critical for its role in supplying energy for cell processes; dubbed the 'molecular currency' or 'fuel' vital to biology.

Adhesion – the clinging together of two or more different materials resulting from attractive forces between their atoms or molecules. Compare with *cohesion*.

Airfoil – an idealized two-dimensional, streamlined shape that produces lift in response to fluid flow. When sliced along the plane that contains the lift vector, a three-dimensional wing typically has the cross-section of an airfoil.

Allen's rule – an observed general trend that animals inhabiting colder regions tend to have shorter, thicker appendages and limbs than otherwise similar individuals in warmer climates. This reduces the surface area over which heat loss occurs.

Alula – a small projection of several feathers on the leading edge of a bird's wing, attaching at the first digit, or 'thumb'. When extended, it plays a role similar to that of an airplane's slat, altering the local airflow and enabling the wing to achieve a higher angle of attack without stalling.

Amplitude – in wave motion, the maximum displacement of a medium or property relative to its average value.

Analog-to-digital converter (ADC) – an electrical circuit that plays a central role in the digitization process by sampling an analog signal at a particular rate and creating a sequential stream of digital values that correspond to the signal's time-varying voltage.

Angular momentum – the rotational analog of momentum, which, for a spinning object, depends on the rate at which it is rotating as well as the mass, shape and distribution of its material. Angular momentum is conserved for an object or system when no external torque acts upon it.

Anther – the part of a flower's stamen that produces and contains pollen, typically located at the end of a long filament.

Aquaporin – a specialized protein channel in biological cell membranes that enables an increase in the transport of water into or out of the cell.

ASA (American Standards Association) – historically, a metric used to quantify a photographic film's speed or sensitivity to light, which has since been replaced by the designation ISO.

Barbs – the hairlike parts of a feather that extend from the central shaft, or rachis, to form the vane.

Barbules – the minute parts of a feather that branch off from, and are smaller than, the barbs. They feature interlocking structures that permit adjacent barbs to remain fixed in place, giving the feather a strong yet flexible character.

Bergmann's rule – an observed general trend that animals inhabiting colder regions are typically larger than otherwise similar individuals in warmer climates. This reflects an increase in volume relative to surface area, which reduces heat loss.

Bernoulli's principle – a law that captures the relationships between pressure, location and speed in a moving fluid, derived from the law of conservation of energy.

Bit depth – the number of bits allocated to represent a digital sample. A larger bit depth requires more resources but enables greater resolution and accuracy. Modern recorders with a depth of 32 bits provide an effectively unlimited dynamic range for audio.

Boids – a simple algorithmic model of flocking bird behavior, invented by Craig Reynolds in 1986, which is capable of producing complex, realistic simulations of murmurations.

Buoyancy – an upward force that a surrounding fluid exerts against the weight of a body, arising from the increase in fluid pressure with depth.

Cardioid microphone – a microphone with nonuniform sensitivity such that, when plotted in two dimensions, it has the 'heart'-like shape of a cardioid curve, with maximum sensitivity directly in front of the device and minimum sensitivity to the rear.

Carotenoids – a class of over 1,000 organic molecules that function as pigments and typically produce red, orange or yellow coloration. Their long sequences of conjugated bonds (see *conjugation*) result in the preferential absorption of short-wavelength light, giving rise to their characteristic colors.

Charge – a property of certain elementary particles that allows them to couple to the electromagnetic field in one of two ways: positive or negative.

Chromatic aberration – an image defect caused by the wavelength dependence of the refractive index of the material(s) comprising a lens system. It manifests as a color-dependent change to the focal length, giving rise to colored fringes. The defect is mitigated through the use of compound lenses with differing refractive indices.

Circle of confusion – an optical spot with a non-zero radius that results from light passing through a lens that is not properly focused. It is used to quantify depth of field (DOF).

Coandă effect – the propensity for a localized stream or 'jet' of fluid to follow a convex surface and entrain the surrounding fluid in the process.

Cochlea – an organ of the inner ear where mechanical energy from small, vibrating bones is transduced into nerve signals. It is coiled in humans, but linear in birds.

Cohesion – the property of a fluid to stay 'bound up' (for example, as in forming a droplet) due to the attractive forces between the molecules composing it. Compare with *adhesion*.

Coma – an image defect or aberration in which off-axis portions of the image become distorted. If the image is a field of stars, for example, those on the periphery will have 'tails' while those in the center are point-like and sharp.

Compression – the application of inward forces at different points in a body or structure, which act to reduce the body's size in one or more directions. The opposite of tension.

Condenser microphone – a transducer that employs a capacitor (historically called a condenser) to convert audio energy into electrical energy. This is accomplished by allowing the changing air pressure to alter the separation of the plates of a charged capacitor, producing a time-varying voltage.

Conjugation – an overlapping of molecular orbitals that allows electrons to become delocalized. Long sequences of conjugated bonds in certain organic molecules enable the absorption of certain wavelengths of visible light. These bonds are a key feature of pigments such as carotenoids and melanins.

Convection – a form of heat transfer through a fluid involving the movement of the fluid itself. This typically occurs due to changes in local density with temperature, which cause pressure differences and currents that mix different portions of the fluid.

Cornea – the transparent front portion of the eye covering the iris and pupil, where most of the refraction of light occurs.

Cosine – for a right triangle, the cosine of an angle is the ratio of its adjacent side to the hypotenuse. When plotting this ratio as a function of angle, the familiar sinusoid wave shape results.

Cotransporter – a complex structure in a cell membrane that facilitates the movement of one type of molecule against its concentration gradient by leveraging the motion of a different molecule with its concentration gradient.

Coulomb's law – a mathematical relationship stating that the force exerted by one point charge on another is proportional to the product of their charges divided by the square of their separation. This is an inverse square law, similar to the law governing gravitational attraction.

Countercurrent exchanger – a device allowing two non-mixing flows, moving in opposite directions, to exploit their proximity in order to exchange some property (such as heat) more efficiently. Networks of blood vessels use such an arrangement to conserve heat (*see rete mirabile and venae comitantes*).

Cryptochrome – a large protein found in avian retinas, thought to supply a mechanism for biological magnetoreception. Exposure to blue light can produce a radical pair within the molecule, with subsequent reactions dependent upon the protein's orientation relative to the earth's magnetic field.

Current – (1) a flow of electric charge through a conductor; the time rate of change of charge. (2) In a fluid, the magnitude and direction of the flow of its molecules.

Decibel (dB) – a unit of power measurement on a logarithmic scale, which is well suited to the expression of relative changes over many orders of magnitude. Human hearing has a dynamic range of 120 dB, which is a factor of one trillion.

Deprotonation – the removal or transfer of a proton from one molecule to another. This process plays an important role in the magnetoreception mechanism of cryptochrome.

Depth of field (DOF) – the maximum difference in distance from an optical device that objects can have while remaining acceptably focused. Increasing aperture decreases the DOF.

Deuteron – a proton bound to a neutron, a key component in the chain of nuclear reactions that power the sun.

Dielectric – an electrically insulating material that becomes polarized by the application of an electric field.

Diffraction – an effect in which waves encountering a barrier or an aperture undergo a deviation in their direction of propagation, such as spreading or bending. This often leads to interference effects.

Diffusion – the movement of particles (or other bodies) with an underlying random character, which in total results in a flow from regions of high concentration to regions of low concentration.

Diopter – a measure of the optical power of a lens, defined as the inverse of its focal length.

Discrete Fourier transform (DFT) – a mathematical procedure for converting a time-based sequence of values (such as voltage amplitudes) into its corresponding frequency-domain expression. This is the central process in the creation of spectrograms.

Dispersion – an effect in which the velocity of a wave through a medium depends on its frequency. In glass, dispersion occurs because the polarizability of the material increases with frequency, which can lead to the separation of white light into a colored spectrum.

Downwash – the air directed generally downward by a wing or airfoil in reaction to the upward lift force generated by the airflow.

Drag – the opposing force exerted by a fluid against the motion of an object moving through it, dependent on its relative velocity.

Electron – a negatively charged fundamental particle, primarily responsible for the chemical behavior of atoms and molecules. An electron is a localized quantum excitation in the electron field that also possesses intrinsic spin and magnetism.

Endotherm – an organism that maintains its body temperature within a certain range via the process of thermoregulation, largely by generating its own heat internally rather than relying on ambient heat.

Energy – the measure of a system's capability to do work, generally classified as either potential energy (which can be stored in a body or system) or kinetic energy (due to motion).

Exit pupil – the effective aperture of light exiting an optical instrument. It is the image of the aperture stop, which is typically the border of the objective element that admits light into the system. The term also denotes the diameter of this aperture, determined for a telescope or binoculars by dividing the objective diameter by the magnification.

Eye relief – the maximum distance between the eye and the eyepiece such that the full field of view (FOV) is visible.

Fast Fourier transform (FFT) – an efficient algorithm for performing a discrete Fourier transform (DFT) or each section of a short-time Fourier transform (STFT). It exploits various symmetries inherent to Fourier analysis to make rapid conversions into the frequency domain.

Field – a physical entity with a value at every point in space. Particular fields, such as the electromagnetic field or the electron field, are fundamental structures in nature, with the quantum excitations in these fields being equivalent to individual photons or electrons, respectively.

Field of view (FOV) – a measure that quantifies the solid angle over which an optical device can accept light, typically expressed as an angular width.

Filter – a device or structure that modifies the harmonic content of a signal passing through it.

Flavin adenine dinucleotide (FAD) – a key protein found within cryptochrome and thought to provide the site at which the absorption of blue light leads to the creation of a radical pair, which is shared across the FAD and a nearby tryptophan molecule. This pair forms the basis for the putative magnetoreception capabilities of some bird species.

Force – that which causes a change in the momentum of a massive body, equal to the time rate of change of momentum. For a constant mass, force is the product of mass and acceleration. It is a vector quantity, always accompanied by an equal and opposite force.

Fourier analysis – the mathematical rules for deconstructing a complex periodic function in terms of a collection, or series, of individual sine waves, known as a Fourier series.

Fovea – a pit-like structure in the retina that features a dense array of cone cells, enabling a level of visual acuity exceeding that which is possible with a flat array of retinal cells. It is critical for the high acuity found in various bird species.

Fractal – a geometrical shape with self-similar properties such that, under increasing magnification, certain key features continue to appear. Fractals are generated by iterating simple arithmetic or random rules, and are frequently found in nature.

Frequency – the rate at which an oscillation or wave completes a cycle.

F-stop – the ratio of focal length divided by aperture for an optical system or lens; lower numbers correspond to more light and less depth of field (DOF). Also called f-ratio or focal ratio.

Galilean telescope – an early telescope design utilizing a positive lens as an objective and a negative lens as an eyepiece. It provides a non-inverted image, but suffers from a narrow field of view and poor eye relief.

Gloger's rule – an observed general trend that animals inhabiting warmer, more humid regions typically have darker coloration than those in colder, drier climates. In birds, this pattern may arise in part because melanin, which darkens feathers, has antibacterial properties that offer protection against feather-degrading bacteria prevalent in tropical environments.

Gradient – a mathematical description of the rate at which a scalar quantity, such as pressure or concentration, changes along any direction.

Gravitation – a fundamental force of attraction acting between bodies that possess mass, obeying an inverse square law.

Hadron – any composite subatomic particle composed of quarks, such as a proton or neutron.

Hann window – a mathematical function typically applied as part of a short-time Fourier transform (STFT) to reduce spectral leakage.

Harmonic – a sinusoidal component of a periodic waveform with a frequency that is an integer multiple of the waveform's lowest, or fundamental, frequency. The relative amplitudes of the harmonics in an audio signal determine its timbre.

Heat – the transfer of energy between two entities with different temperatures via conduction, convection or radiation.

Hydrophobic – a term referring to a material or molecule that is repelled by water. A hydrophobic surface provides little or no adherence, resulting in the formation of droplets.

Hyperopia – also known as far sightedness; the condition of an eye that can achieve focus for distant objects but not nearby ones, due to insufficient accommodation by the lens.

Image – the representation of an object formed by an optical component or system. If the representation can be projected onto a screen, it is termed a real image; if it appears to come from a location where it is not (and where a screen cannot be placed), it is termed a virtual image.

Inertia – the propensity for matter to resist a change in motion, either remaining at rest or maintaining its velocity.

Infrared – electromagnetic radiation that cannot be detected by the human eye, with a wavelength longer than that of red light but shorter than those of microwaves. Infrared is effective for the transmission of heat by radiation.

Inhibitor – a substance that acts to prevent or decrease the rate of a chemical reaction.

Intensity – the rate at which radiant energy is transferred per unit area.

Interference – a result of the linear superposition of two or more waves in the same spatial location, where the total amplitude may be increased (constructive interference) or decreased (destructive interference) depending on the relative phase relationships of the superposed waves.

Inverse square law – a mathematical relationship in which a squared quantity appears in the denominator; it occurs in various expressions concerning force and intensity.

Ion – an atom or molecule that has a net non-zero charge due to having lost or gained one or more electrons.

Iridescence – the change in color of a surface as the angle at which it is viewed is changed. This is an interference effect that requires a specific structure, such as a thin film, measuring roughly on the order of the wavelength of visible light.

ISO – a numerical value that captures the sensitivity (or speed) of photographic film and digital camera sensors. ISO stands for 'International Organization for Standardization' and was preceded by ASA. With shutter speed and f-stop, it is one of the three key photographic settings.

Isomers – molecules that have the same molecular formula but differing spatial arrangements of their component atoms.

Keplerian telescope – an optical instrument that uses positive (converging) lenses for both the objective and the eyepiece, resulting in improved eye relief and a larger field of view compared to the Galilean telescope, but at the cost of an inverted image.

Keratin – a ubiquitous type of protein found in feathers (as well as in nails, scales and other features) that forms strong, fibrous structures.

Lens – a curved piece of transparent material, such as glass, which causes the controlled deviation of light due to refraction. A positive or converging lens brings parallel rays to a real focal point, while a negative or diverging lens leads to a virtual focal point.

Lift – a force exerted by a fluid on a moving body (typically a wing or airfoil) that is perpendicular to the direction of the oncoming flow. In the case of terrestrial flight, the fluid is air and the lift acts to counter the weight of the body.

Light – an electromagnetic wave with a wavelength between 380 to 750 nanometers, which is the range that is perceptible by human vision.

Loudness – the physiological perception of the intensity of sound.

Magnet – an object that produces a magnetic field, which is closely allied with the electric field, together comprising the fundamental electromagnetic field. All magnets possess two poles, termed north and south. They may be made from materials with intrinsic magnetism or as a device that utilizes electrical currents, known as an electromagnet.

Magnetoreception – the sensory perception by an organism of the earth's magnetic field, which may be used to aid migration. Members of various invertebrate and vertebrate families, including birds, are known to possess this sense.

Magnification – a dimensionless measure of the degree to which an optical component or system alters the size of an image.

Mass – a measure of an object's resistance to change in motion, or inertia, determined by the quantity of material making up the body.

Melanins – a broad category of complex biological molecules that function as dark pigments due to their proclivity to absorb wide portions of the visible spectrum. The key components giving black and brown feathers their color, melanins also provide UV protection, structural strength and antimicrobial properties.

Melanosome – a cell organelle that produces and contains melanin.

MEMS (micro-electromechanical systems) – device technology at a microscopic scale (typically from 1 to 100 microns) that involves both electronic and moving parts. Condenser microphones in modern cellphones are an example.

Microstructure – the structure of a material at a very small scale, typically in the range of tens of microns or smaller, revealed using optical instruments.

Mie scattering – deflection of electromagnetic radiation by a spherical surface that is comparable in size to, or larger than, its wavelength. Unlike Rayleigh scattering, Mie scattering does not depend on frequency and does not produce color effects.

Momentum – a vector quantity, the product of an object's mass and velocity, which remains constant unless acted upon by an external force.

Murmuration – a large flock of birds that appears to move in an elegantly coordinated manner, but which is merely governed by simple local rules that produce emergent larger-scale effects.

Navier–Stokes equation – the general equation that describes the behavior of typical fluids.

Neutron – a neutral hadron composed of one up quark and two down quarks, held together by gluons. Together with protons, neutrons comprise atomic nuclei.

Noether's theorem – a fundamental rule in mathematical physics stating that a continuous symmetry must correspond to a conserved quantity. For example, rotational invariance leads to the conservation of angular momentum.

Nyquist's theorem – a fundamental rule of digital signal processing specifying that the minimum sampling rate must be at least twice the largest frequency to be captured.

Opsin – a type of protein integral to the light-detection capability of rod and cone cells, where it is attached to a light-sensitive molecule such as retinal.

Orbital – a spatial region in an atom or molecule where an electron with a particular quantum state may be found.

Oscillation – a complete cycle of periodic movement relative to an equilibrium position.

Particle – a basic constituent of matter or a carrier of a fundamental force, corresponding to a single excitation in a quantum field.

Pauli exclusion principle – a foundational rule of physics and chemistry stating that two or more identical matter particles (such as electrons or quarks) cannot share the same quantum state while in proximity (e.g., within the same atom).

Phase – (1) a measure indicating a particular fraction, point or stage within a wave's cycle, measured in radians or degrees. The difference in phase between two otherwise identical waves determines the character of their interference. (2) The state of a sample of matter (e.g., solid, liquid, gas). (3) In some autofocus systems, the lateral displacement of an image resulting from an off-axis restriction of the aperture.

Photodiode – a junction made of positively and negatively doped semiconductors that generates a current when illuminated with light. This is the basic device that enables digital photography.

Photon – a massless particle corresponding to a single quantum excitation in the electromagnetic field, which travels at the speed of light.

Pigment – a large class of molecules that preferentially absorb certain wavelength ranges within the visible spectrum, resulting in the production of color upon reflection. See *carotenoids* and *melanins*.

Polarizability – a property of materials, such as dielectrics, in which charge is redistributed when exposed to an electric field without resulting in the flow of a continuous current.

Polarization – the orientation of the oscillatory motion of a transverse wave along a direction perpendicular to its propagation. In light, it refers to the configuration of the oscillating electric field.

Porro prism – an optical device that uses four total internal reflections to produce a left–right and up–down image reversal while laterally displacing the beam. Often used in binoculars, it requires only anti-reflection coating.

Potential – a scalar quantity associated with a field, such that the gradient of the potential gives the force.

Power – the time rate of change at which energy is delivered or work is performed.

Precession – a persistent change in the orientation of a rotating body or a particle with angular momentum or spin. It is typically caused by the application of a torque, such as gravity acting on a spinning top or a magnetic field acting on an electron.

Pressure – the force acting perpendicularly against a surface divided by the area over which it acts.

Proton – a stable, positively charged hadron composed of two up quarks and one down quark, held together by gluons. Along with neutrons, protons make up atomic nuclei.

Quantum – a fixed, minimal amount by which certain properties, such as energy, can change. This is in contrast to properties that change continuously. These fixed amounts are fundamental to quantum theories that describe the behavior of molecules, atoms, light and subatomic particles.

Quark – an elementary particle that combines with others to form larger composite particles (hadrons), such as protons and neutrons.

Rachis – the central shaft of a feather, from which the barbs extend.

Radiation – the motion of particles or waves emitted from one or more sources, propagating across space; derived from the Latin word for the spoke of a wheel.

Radical pair – two valence electrons that were previously paired but have been separated, typically as unpaired valence electrons on adjacent molecules. This separation enables a mechanism by which an external magnetic field (such as the earth's magnetic field) can influence the subsequent chemistry, which is thought to play a role in magnetoreception through cryptochromes.

Rarefaction – a region in a fluid where pressure and density are lower than in adjacent areas. Rarefaction alternates with regions of compression when sound waves are produced.

Rayleigh scattering – the deflection of light due to its interaction with atoms, molecules or particles that are significantly smaller than the light's wavelength. Since the response is dependent upon the polarizability of bound electrons, which increases with frequency, the net effect on the visible spectrum is to scatter bluish light preferentially.

Reflection – a change in the direction of particles or waves upon encountering, but not crossing, the boundary between different media.

Refraction – a change in a wave's direction of propagation as it passes from one medium to another, caused by the change in speed (and wavelength). Refraction occurs when a wave strikes a boundary at an angle other than perpendicular and is fundamental to the operation of optical devices like lenses.

Refractive index – a dimensionless number that characterizes the speed of light in a material, defined as the ratio of the speed of light in a vacuum to the speed of light in the material. The refractive index varies with the frequency of the light, which leads to dispersion.

Resistance – the measure of an electrical element's opposition to the flow of charge, or current.

Resolution – the minimum distance (or angle) between light sources or features such that they can be discerned through an optical instrument.

Resonance – the oscillation of a system or body, often with large amplitude, when it is driven at its natural frequency of vibration.

Rete mirabile – a network of adjacent veins and arteries that functions as a countercurrent exchanger, aiding in thermoregulation.

Scalar – a quantity that can be expressed with a single value, as opposed to a vector, which involves both a magnitude and a direction.

Schmidt–Pechan prism – a type of compound roof prism with six internal reflections, commonly used in binoculars to provide an erect, non-reversed image without deviating the light off the optical axis, unlike the Porro prism. It is more compact than the Abbe–König prism, but requires an additional dielectric mirror coating, along with anti-reflection and phase coatings.

Semiconductor – a crystalline solid material with an intermediate electrical conductivity, which can be tailored by introducing impurities. Various devices for controlling electrical currents and/or sensing light can be constructed from the junctions of different semiconductors. Such structures are highly amenable to reductions in size, leading to enormous improvements in digital electronics, sensing and computation.

Short-time Fourier transform (STFT) – the principal method used to change a long, evolving time-domain signal into a sequential series of corresponding Fourier transforms. It requires the use of a window function (see *Hann window*) to reduce spectral leakage. The STFT is integral to the creation of spectrograms from birdsong audio and typically uses the FFT algorithm.

Sine – for a right triangle, the sine of an angle is the ratio of its opposite side to the hypotenuse; when plotting this ratio as a function of angle, the familiar sinusoid wave shape results.

Sinusoid – also termed a sine wave, a periodic wave with a shape determined by the trigonometric sine function. It is the fundamental unit by which a Fourier series can construct a periodic waveform.

Snell's law – the mathematical relationship describing the angular deviation of a light beam crossing a boundary between materials with different refractive indices.

Spectral leakage – a distortion to the frequency spectrum that occurs when a signal is truncated during the course of a short-time Fourier transform (STFT), causing energy to leak into nearby frequencies. It can be mitigated by introducing a window such as the Hann window.

Spectrogram – a visual representation of the time evolution of the frequencies present in a signal, especially useful for analyzing bird vocalizations. It is constructed by making a color-coded or gray-scale map of the results of a short-time Fourier transform (STFT) of an audio signal versus time.

Spectrum – a property that varies over a range of frequencies (or wavelengths) for which some wave phenomena are of interest. It typically refers to electromagnetic or audio waves.

Speed – the scalar quantity that represents the rate at which an object changes its position over time, calculated as the magnitude of the velocity.

Spherical aberration – an image defect that occurs when different radial portions of a lens do not focus to the same point. It is an inherent property of spherical lenses.

Spin – an intrinsic property of many fundamental particles (and their corresponding quantum fields), which confers a small, fixed amount of angular momentum. Charged particles with spin also have an intrinsic magnetic character, resulting in their precession when subjected to a magnetic field.

Stall – the loss of lift for a wing or airfoil, which occurs when the angle of attack exceeds a critical angle.

Standard Model – a comprehensive theory that describes all known matter particles and three of their interactions (electromagnetic, weak and strong). It is based on quantum field theories and capable of producing the most accurate experimental predictions ever confirmed.

Superposition – the principle that the net amplitude for any number of waves at any point in space is found by simply adding the individual amplitudes for each wave.

Symmetry – a property of an object or system such that a change, such as a rotation, reflection or translation, has no net effect.

Syrinx – the vocal organ of birds, which uses vibrating membranes whose tension (and hence frequency) is controlled by muscles held by a series of rigid rings. Constricted air flow is used to initiate the vibrations, while the shapes of the syrinx, throat and mouth further act to amplify and filter the sound.

Tangent – for a right triangle, the tangent of an angle is the ratio of the opposite side to the adjacent side.

Temperature – the property of an object or spatial region that determines whether there is a net flow of heat in from or out to some other object or region in contact with it. Having a lower temperature indicates that heat will flow in; a higher temperature indicates that heat will flow out.

Tension – a force transmitted via pulling or stretching along an extended body. The opposite of compression.

Thermoregulation – the process by which an organism maintains its body temperature within an optimal range, even when the surrounding temperature differs significantly.

Thrust – a reaction force resulting from the displacement of mass in the opposite direction. In the case of bird flight, thrust is the propelling, forward force generated from a full wingbeat cycle that is necessary to maintain sufficient airspeed in the presence of drag.

Timbre – the quality or 'color' of an audio tone, which distinguishes different kinds of instruments sounding the same note. It is determined by the specific harmonic content of the signal.

Time domain – the expression of a function with time as the independent variable. By performing a Fourier transform on a time domain signal, the corresponding frequency domain signal is determined.

Total internal reflection – a reflection in which no energy is transmitted across the boundary. In optics, this occurs when light moving in a slower (higher index) medium arrives at a sufficiently oblique angle at a boundary with a faster medium such that Snell's law has no real-valued solutions. This is utilized in prisms.

Transducer – any device that enables one form of energy to be converted into another. It is typically used to accurately change a time-varying signal into a more useful version. For example, a microphone transduces an audio signal into an electrical signal that is amenable to recording.

Tryptophan – an amino acid found within cryptochrome. Together with the FAD molecule, it plays a critical role in facilitating the radical pair thought to enable magnetoreception in birds.

Turbulence – a disordered flow of a fluid in which the individual particles comprising it move in irregular paths, with momentum transferred from one region to another.

Upwash – an upward deflection of air resulting from pressure differences as the air flows past a wing. It may be especially pronounced near the wingtips and is utilized by bird flocks that take on the familiar V-shape.

Valence electrons – the electrons in the outermost (highest energy) shell of an atom, which can participate in the formation of chemical bonds.

Vector – a mathematical quantity with both magnitude and direction. Many important physical properties are vectors, including velocity, force, angular momentum and drag.

Velocity – the time rate of change of position; a vector quantity with both magnitude (speed) and direction.

Venae comitantes – a biological countercurrent exchanger consisting of one or two veins lying adjacent to an artery, and important for thermoregulation.

Viscosity – a fluid's resistance to flow.

Voltage – also known as electromotive force, or electrical potential; equal to the electrical potential energy divided by the charge.

Vortex – a flow in a fluid that has a non-zero rotation, or curl; a localized region in which the fluid revolves around a line. *Vortices* (plural) are induced by the motion of a wing through the air.

Wave – a periodic disturbance with some spatial extent, and characterized by frequency, wavelength, amplitude and velocity. A wave with non-zero velocity transfers energy from one region to another.

Wavelength – the spatial analog to the period of a wave. It is the distance at which a wave displays the identical shape or form. Typically measured as the distance between consecutive crests or troughs, or any successive points with the same phase.

Wavenumber – the spatial analog to the frequency of a wave. Typically expressed as 2π divided by the wavelength.

Weight – the force exerted on a mass by gravity.

Wing loading – the ratio of the weight of a bird (or any flying animal or aircraft) to the total wing area.

Winglet – a small vertical panel attached to the distal end of a wing, which reduces wingtip vortices.

Work – a scalar quantity equivalent to the energy delivered by a force acting on a body that is thereby displaced. It is determined by the product of the distance moved and the component of the force in that direction.

Bibliography

Alberts, B., Bray, D., Hopkin, K. et al. (2014) *Essential Cell Biology*, 4th edition. New York, NY: Garland Science
Anderson, D.F. and Eberhardt, S. (2001) *Understanding Flight*. New York, NY: McGraw-Hill
Anderson, J.D. (2011) *Fundamentals of Aerodynamics*, 5th edition. New York, NY: McGraw-Hill
Asimov, I. (1993) *Understanding Physics*. Dorchester: Dorset Press
Barnsley, M.F. (1998) *Fractals Everywhere*, 2nd edition. Boston, MA: Academic Press Professional
Bech, C. and Reinertsen, R.E. (eds) (1989) *Physiology of Cold Adaptation in Birds*. NATO ASI Series, 173. Boston, MA: Springer
Bird, D.M. (2004) *The Bird Almanac: A Guide to Essential Facts and Figures of the World's Birds*. Buffalo, NY: Firefly Books
Bruice, P.Y. (2013) *Organic Chemistry*, 7th edition. London: Pearson
Carlson, J., Jaffe, A. and Wiles, A. (eds) (2006) *The Millennium Prize Problems*. Providence, RI: American Mathematical Society
Carlson, W.E. (2017) *Computer Graphics and Computer Animation: A Retrospective Overview*. https://books.google.com/books?id=kpnxswEACAAJ
Cleere, N. (1998) *Nightjars: A Guide to the Nightjars, Nighthawks, and Their Relatives*. New Haven, CT: Yale University Press
Cropper, W.H. (2001) *Great Physicists*. New York: Oxford
Davis, R.E., Gailey, K.D. and Whitten, K.W. (1984) *Principles of Chemistry*. Philadelphia, PA: Saunders College Publishing
Dodds, W.K. and Whiles, M.R. (2020) *Freshwater Ecology: Concepts and Environmental Applications of Limnology*. London: Academic Press
Dunn, R. (2011) *The Telescope: A Short History*. London: Conway
Ehrlich, P., Dobkin, D.S. and Wheye, D. (1988) *The Birder's Handbook: A Field Guide to the Natural History of North American Birds*. New York, NY: Simon and Schuster
Einstein, A., Furth, R. and Cowper, A.D. (1926) *Investigations on the Theory of The Brownian Movement*. London: Methuen & Co.
Einstein, A. and Infeld, L. (1938) *The Evolution of Physics*. Cambridge: Cambridge University Press

Eisberg, R. and Resnick, R. (1985) *Quantum Physics of Atoms, Molecules, Solids, Nuclei, and Particles*, 2nd edition. New York, NY: John Wiley & Sons

Erbe, C. and Thomas, J.A. (eds) (2022) *Exploring Animal Behavior Through Sound: Volume 1 Methods*. Cham, Switzerland: Springer Nature.

Feynman, R.P. (1988) *QED: The Strange Theory of Light and Matter*. Princeton, NJ: Princeton University Press

Feynman, R.P., Leighton, R.B. and Sands, M. (2011) *The Feynman Lectures on Physics*, New Millennium edition. New York, NY: Basic Books

Feynman, R.P. and Robbins, J. (2005) *The Pleasure of Finding Things Out: The Best Short Works of Richard P. Feynman*. New York, NY: Basic Books

Fletcher, N.H. and Rossing, T.D. (1991) *The Physics of Musical Instruments*. New York, NY: Springer-Verlag

Fox, M. (2010) *Optical Properties of Solids*, 2nd edition. Oxford: Oxford University Press

Franklin, B. (1751) *Experiments and Observations on Electricity, Made at Philadelphia in America, by Mr. Benjamin Franklin, and Communicated in several Letters to Mr. P. Collinson, of London, F.R.S.* London: Printed and sold by E. Cave, at St. John's Gate

Fraknoi, A., Morrison, D. and Wolff, S. (2022) *Astronomy*, 2nd edition. Houston, TX: OpenStax

Gell-Mann, M. (1994) *The Quark and the Jaguar: Adventures in the Simple and the Complex*. New York, NY: W.H. Freeman and Co.

Gleick J. (1987) *Chaos: The Making of a New Science*. New York, NY: Viking

Gleick, J. (2004) *Isaac Newton*. New York, NY: Vintage

Golub, L. and Pasachoff, J.M. (2014) *Nearest Star: The Surprising Science of our Sun*. Cambridge: Cambridge University Press

Gould, S.J. (1996) *Full House: The Spread of Excellence from Plato to Darwin*. New York, NY: Harmony Books

Greenewalt, C.H. (1991) *Hummingbirds*. Mineola, NY: Dover Publications

Griffiths, D. (1987) *Introduction to Elementary Particles*, 2nd edition. New York, NY: John Wiley & Sons, Inc.

Hammond, J.H. (1981) *The Camera Obscura: A Chronicle*. Boca Raton, FL: CRC Press

Hecht, E. (2017) *Optics*, 5th edition. Harlow: Pearson Education Limited

Horváth, G. (ed.) (2014) *Polarized Light and Polarization Vision in Animal Sciences*. New York, NY: Springer

Horváth, G. and Varjú, D. (2004) *Polarized Light In Animal Vision: Polarization Patterns In Nature*. New York, NY: Springer

Houghton, E.L., Carpenter, P.W., Collicott, S.H. et al. (2012) *Aerodynamics for Engineering Students*, 6th edition. Oxford: Butterworth-Heinemann

Johnsgard, P.A. (1983) *The Hummingbirds of North America*. Washington, DC: Smithsonian Institution Press

Johnson, C., Jr. (2017) *Science for the Curious Photographer: An Introduction to the Science of Photography*, 2nd edition. New York, NY: Routledge

Johnson, G. (1999) *Strange Beauty: Murray Gell-Mann and the Revolution in Twentieth-Century Physics*. New York, NY: Knopf

Kundu, P.K. and Cohen, I.M. (2007) *Fluid Mechanics*, 4th edition. Burlington, MA: Academic Press

Langton, C., Levin, V., Tishman, R. *et al.* (2017) *The Intuitive Guide to Fourier Analysis & Spectral Estimation*. Chicago, IL: Mountcastle Academic

Lautrup, B. (2011) *Physics of Continuous Matter*, 2nd edition. Boca Raton: CRC Press

Lederman, L. and Teresi, D. (2006) *The God Particle: If the Universe Is the Answer, What Is the Question?* New York, NY: Houghton Mifflin

Lockhart, P. (2009) *A Mathematician's Lament*. New York, NY: Bellevue Literary Press

Lowrie, W. (2007) *Fundamentals of Geophysics*. Cambridge: Cambridge University Press

Lynch, D.K. and Livingston, W. (2001) *Color and Light in Nature*, 2nd edition Cambridge: Cambridge University Press

Maloberti, F. and Davies, A.C. (eds) (2016) *A Short History of Circuits and Systems*. Gistrup, Denmark: River Publishers

Marche, S. (2012) *How Shakespeare Changed Everything*. New York, NY: Harper Perennial

Marieb, E.N. and Keller, S.M. (2021) *Essentials of Human Anatomy and Physiology*, 13th edition. Harlow: Pearson

Merlitz, H. (2023) *The Binocular Handbook: Function, Performance and Evaluation of Binoculars*. Cham, Switzerland: Springer

Mindlin, G.B. and Laje, R. (2005) *The Physics of Birdsong (Biological and Medical Physics, Biomedical Engineering)*. New York, NY: Springer

Oerter, R. (2006) *The Theory of Almost Everything: The Standard Model, the Unsung Triumph of Modern Physics*. New York, NY: Plume

Paulos, J.A. (2001) *Innumeracy: Mathematical Illiteracy and Its Consequences*. New York, NY: Holt McDougal

Pennycuick, C.J. (2008) *Modelling the Flying Bird*. Boston, MA: Academic Press

Pepperberg, I.M. (2002) *The Alex Studies: Cognitive and Communicative Abilities of Grey Parrots*, 1st edition. Cambridge, MA: Harvard University Press

Poe, E.A. (1848) *Eureka: A Prose Poem*. New York, NY: Geo. P. Putnam

Proctor, N.S. and Lynch, P.J. (1993) *Manual of Ornithology: Avian Structure and Function*, 1st edition. New Haven, CT: Yale University Press

Quain, J., Schafer, E.A., Sharpey, W. *et al.* (1878) *Quain's Elements of Anatomy*, 8th edition. New York, NY: William Wood

Rayburn, R.A. (2012) *Eargle's The Microphone Book*, 3rd edition. Burlington, MA: Focal Press

Read, O. and Welch, W.L. (1976) *From Tin Foil to Stereo: Evolution of the Phonograph*. Indianapolis, IN: Howard W. Sams

Riley, P.A. and Borovansky, J. (2011) *Melanins and Melanosomes: Biosynthesis, Structure, Physiological and Pathological Functions*. Weinheim, Germany: John Wiley & Sons

Rovelli, C. (2016) *Seven Brief Lessons on Physics*. New York, NY: Riverhead Books

Sagan, C. (1997) *The Demon-Haunted World: Science as a Candle in the Dark*. New York: Ballentine Books
Sarangan, A. (2020) *Optical Thin Film Design*, 1st edition. Boca Raton, FL: CRC Press
Scanes, C.G. (2015) *Sturkie's Avian Physiology*, 6th edition. London: Academic Press
Schumm, B.A. (2004) *Deep Down Things: The Breathtaking Beauty of Particle Physics*. Baltimore, MD: Johns Hopkins University Press
Seeds, M.A. and Backman, D.E. (2016) *Foundations of Astronomy*, 13th edition. Boston, MA: Cengage Learning
Serway, R.A. and Jewett, J.W. (2010) *Physics for Scientists and Engineers*, 8th edition. Belmont, CA: Brooks/Cole
Sigee, D.C. (2005) *Freshwater Microbiology*. West Sussex: John Wiley and Sons
Smith, S.W. (1999) *The Scientist and Engineer's Guide to Digital Signal Processing*. San Diego, CA: California Technical Publishing
Sprawls, P. (2001) *Magnetic Resonance Imaging: Principles, Methods, and Techniques*. Madison, WI: Medical Physics Publishing
Stefanov, K.D. (2022) *CMOS Image Sensors*. Bristol: IOP Publishing
Stull, R. (2015) *Practical Meteorology: An Algebra-based Survey of Atmospheric Science*. Madison, WI: Sundog Publishing
Tegmark, M. (2015) *Our Mathematical Universe: My Quest for the Ultimate Nature of Reality*. New York, NY: Vintage
Tennekes, H. (2009) *The Simple Science of Flight*. Cambridge, MA: MIT Press
Voet, D. and Voet, J.G. (2011) *Biochemistry*, 4th edition. New York, NY: Wiley
Walker, J. (1978) *The Flying Circus of Physics With Answers*. New York, NY: John Wiley & Sons, Inc
Wilson, B.W. (1980) *Birds: Readings from Scientific American*. San Francisco, CA: WH Freeman & Co
Wood, C.A. (1917) *The Fundus Oculi of Birds, especially as viewed by the Ophthalmoscope: A Study in Comparative Anatomy and Physiology*. Chicago, IL: Lakeside Press
Wulf, A. (2016) *The Invention of Nature: Alexander von Humboldt's New World*. New York, NY: Vintage

Index

Page numbers in **bold** refer to figures.

Abbe numbers 163, 164, **164**
acceleration 25, 68, 183, 188, 197, 221, 225
Acrocephalus scirpaceus 32
action at a distance 27, 30, 32, 56
active transporters 43–4, **44**
adenosine diphosphate 45
adenosine triphosphate 44, 45, 221
adhesion 29, 199, 221
Aegolius acadicus 84, **84**
African Gray Parrot 11
air 73–4, 175–6
　compression of 74, **75**
　density 74, 189, 197
　pressure 73–80, 82, 85, 183, 185–7, 190, 223
　rarefaction of 74, **75**, 82, 230
　resistance 29, 66, 181, 194, 218 *see also* drag
　velocity 32, 58, 76, 78, 185–8, 190, 192
airfoils **184**, 184–5, **186**, 186, **187**, 187, 189–90,
　　190–1, 191, **193**, 219, 221, 231
albatrosses 48, 198
　Black-browed Albatross 49
　Snowy Albatross 48, 193
Alectoris rufa 21
algorithms 18, 49, 97, 222, 225, 230 *see also* boid
　simulations
Allen's rule 173, 180, 221
Ammospiza caudacuta 173
analog-to-digital converter 89, 142, 221–2
Anarhynchus frontalis 23
Anas platyrhynchos 197
angular momentum 22, 23, 222, 231
　conservation of 187, 198, 222, 228
animals
　adaptations to environment 173–4, 178–81,
　　221, 226

patterns and colors 51–2 *see also* birds,
　physiology of: feathers
aperture 87–8, **136**, 137–8, **138–9**, 141, **141**, 142,
　144, 146, 165, **167**, 167, 209, 224,
　226, 229
aperture stop 154–5, **155**, 157, 226
Aptenodytes forsteri 146
aquaporins 43, **44**, 44–5, 222
Aquila chrysaetos 49
Arctic Tern 37
arteriovenous anastomosis 174
astronomy 2–4
atomic nuclei 17, 56, 59, 61, 63, 59, 99, 113, 126,
　131, 134, 228
atoms 29, 35, 41, 55, 56, 57, 59, 60, 115, 125–7,
　130, 133, 134, 162, 175, 206, 225, 227
　see also ions; subatomic particles
audio *see also* sound
　dynamic range 90, 142–3, 222, 224
　frequencies 74–5, 87–8, 90, 91–3, **94–5**, 95–6
　signals 79, 80, 82–4, 86–7, **88**, 88–9, **89**,
　　90–1, 92–4, 117, 221–2, 226, 231
　tone 76, 79, 232
　wavelengths 75, **78**, 78, 87, 88, **92–3**, 93, 95

Barnsley ferns 18, 23, 53
Bearded Bellbird 81
bee 'dances' 104
Bergmann's rule 173, 222
Bernoulli's principle 77, 185, 222
binary 90
binoculars 158–60, 165–6, 221, 225, 229, 230
biochemistry 51
biological magnetoreception 207, 224, 225, 227,
　229, 233

biology 30, 133, 171
bird flocks 54, 188, 228, 232
birding 5, 10
birds
  aquatic 177–9
  diving 67–8, 147, 221
  enumeration of 11, 12–13
  identification of 10
  marine 198
  nocturnal 83, 146, 215
birds, behavior of
  communication 80, 82
  flight 20–1, 25–6, **26**, 27, 180, 182, 188, 189, 191–7, 232 *see also* flight, principles of
    flapping 193–5, 196–7
    gliding 193
    hovering 195–6, **196**
    soaring 184, 218
    wingbeat cycle 25–6, **26**, 193–5, 232
    wingbeat frequency 219, 196–7, **197**, 219
  flocking 53, 54 *see also* murmurations
    simulations of 53–4
  foraging 48–9, 109, 199
  hunting 83, 109, 111
  use of to solve optimization problems 49
  vocalizations 76, 79–80, 81, 82, 84, 94, 97, 231
birds, physiology of
  acoustic adaptations 84
  adaptations to environment 169, 172–4, 178–80
  asymmetry 23, 84
  bill shapes 68, 172–4, **199**, 199
  capillary ratchet **199**, 199
  cryptochromes 61–4, 104, 224, 225, 229, 232
  ears 82–4
    cochlea 82–3, 223
  eyes 64, 143–8, 175, 214, 215 *see also* eyes
  feathers **116**, 116–17, 119, 124–5, 176, 177–9, **194**, 194–5, 217
    alula 191, 221
    barbs **116**, 116, **117**, 124, 176, 177–8, 222
    barbules **116**, 116, **124**, 124, 176, 222
    colors 51, 105, 116–17, 119, 179–81, 188, 198
    patterning 179
    rachis **116**, 116, 194, 229
    vanes 116, **194**, 195, 222
  hearing 82–4, 146
  legs 174

  muscles 193–4, 196
  navigation 64, 104
  plumage 21, 51–2, 119, 172, 176, 179–80, 198
    see also birds, physiology of: feathers
  respiratory system 81
  salt glands 42–3, **43–4**, 44–6, 175, 198
  sensing of magnetic fields 61–4, 104, 227 *see also* biological magnetoreception
  sensing of polarized light 104
  sexual dimorphism 111
  syrinxes 76–7, **77**, 79, 80, 231
  thermoregulation 169, 171–5, 179–81
  vision 20–1, 111, 143–8, 199
    low-light 146–7
    underwater 147–8
  wings 26, 181, 188–9, 191–4, **194**, 195–6, 221
    see also wing shape
    bones of 180, 194–6
    morphology 17, 181, 184, 189, 193
bird song, recording of 85, 97, 232 *see also* audio; microphones; sound
bit depth 90, 142, 222
bits 90–1, 222
Black Bustard **53**
Black-capped Chickadee 94
Black Kite **145**
boid simulations 53–4, **54**, 222
Brownian bridge 47
Brownian motion 35, 36, 40, 46
Brown, Robert 34–5, 30
Bubulcus ibis 109
Buteo lagopus 111
Buteo platypterus **144**

California Condor 57
Calonectris
  borealis 49
  diomedea 49
Calypte anna 29
cameras 137, 139–43, 154, 165, 166, **167**, 167–8
  see also lenses; photography
Campephilus principalis 69, 85
capacitors 85, **86**, 223
Capuchinbird 81
carbon 126–7, 133, 206
Carrion Crow 11
Cattle Egret 109
cell membranes 42, 43

transport across 42, 43–4, 83, 132, 222, 223
  see also active transporters; diffusion channels
cells, organic 42, **44**, 44 *see also* eyes
  epithelial 43–4
charges, electrical 27–30, 45, 58, 99, 134, 176, 203, 222
  unbalanced 62
chemical elements 17–18
chemistry 41, 51, 60, 62, 65, 132, 171, 229
  equations 35
  reactions 16, 21, 35, 51–2, 53, 99, 170, **171**, 227
chickadees 80
chromatic aberration 162–3, **163**, 164, 222–3
circles **14**, 14–15, 23
circles of confusion **141**, 142, 223
Clarkia pulchella 33–4, **34**, 40
Clerk Maxwell, James 33
clouds 114–15
Coandă effect 185, 218, 223
cohesion 53–4, 176–8, 199, 223
color 105, 112–17, 119, 122–5, 128–9, 132, 135, 136, 163, 165, 181 *see also* pigments; spectra, colored
  experience of 112, 211
  iridescence 120–5, 161, 227
  structural 115–23, 129
  wavelengths 112, 114, 128
coma 165, 223
Common Kestrel 111
Common Raven 169
compass, chemical 64
compression 67, 74, **75**, 223, 230
conduction, thermal 172, 173, 175, 180, 226
conductivity, heat 170, 174, 175, 180
conic sections **14**, 87, 165 *see also* circles; ellipses; parabolas
conjugated bonds **127**, 128–9, 131, 132, 213, 223
conjugation 127, 223
conservation laws 21–2, 23, 58, 107–8, 185
convection 30, 172, 175, 223, 226
Coracias benghalensis 119
Corvus corax 169
Corvus corone 11
cosine 93, 223
cotransporters 44, 45, 223
Coulomb's law 28, 29, 223
countercurrent exchange **174**, 175

countercurrent exchanger 174, 224, 230, 233
counting 11, *see also* mathematics; numbers
  in animals 11
  of birds 10–11, 12
Crested Serpent Eagle **53**
crossbills 23

deceleration 68–9
dehydration 42–3
density 52, 66, 67, 69, 74–5, **75**, 117, 181, 184–5, 188–9, 197, 223, 230
deprotonation 62–3, 64, 224
dielectrics 161, 224, 229, 230
diffraction 87, 146, 209, 224
diffusion 42, 46, 51–2, 224
diffusion channels 42–3, **43**, 46, 51–2
Diomedea exulans 48, 193
diopters 148, 224
direction 23, 25, 32, 46–7, **47**, 54, **62**, 64, 71, 73–4, 84, 86–8, 102–5, 110, 114, 118, 136–8, 144, 150, 152, 158, **184**, 185, 192–3, 195, 221, 224, 224, 231, 230
distance 22, 24, 26, 28, 38–40, 46–7, 56, 58–9, 67, 70–1, 75, 117, 118, 122, 139–41, 152, 156, 177, 186, 192–3, 209–10
Domestic Chicken 11
doves 80
drag 21, 66–7, **184**, 184, 190–2, **192–3**, 193, 198, 219, 224, 232

earth, the 31 *see also* geomagnetism
  orbit 14, **15**, 15–16
Eastern Bluebird 50
Eastern Whipbird 81
echolocation 80
eggs **17**, 17
egrets 105, 109–10, 191
Egretta gularis 109
Einstein, Albert 22, 35, 40, 46
electricity
  conductivity 133, 230 *see also* semiconductors
  current 30–2, 88–9, 127, 133–4, 135, 223
  electrical repulsion 28, 56, 99
  voltage 30, 85–6, **86**, 89, 142–3, 222, 223, 233
electromagnetism *see* fields, electromagnetic
electron field *see* fields, quantum
electronics 86, 88, 90, 135 *see also* microphones; sound, recording technology

light sensors 133, **135**, 135–6, **136**, 137, 140, 141, **141**, 142–3, **167**, 167–8, 213, 227
photodiodes 134–5, **135**, 136, 142, 143, 229
electron pairs
　orbital 61–2, **62**
　radical 61–2, **62**, 63–4, 104, 207, 224, 225, 229, 232
ellipses **14**, 14–15, 17, 27, 165
emergence 54
endotherms 169–70, 171, 225
energy 22, 130, 131–2, 225
　acoustical 82
　conservation of 77, 172, 185, 222
　conversion of 82, 85, 125, 134–5, 221, 232
　　see also transducers
　kinetic 22, 128, 170, 175, 225
　mechanical 82
　potential 22, 30, 44, 45, 85
　solar 13–14, 30
　thermal 170–2 *see also* heat
Eupodotis atra etoschae **53**
Eurasian Reed Warblers 32
Eurasian Woodcock 117
evaporation 172, 216
exit pupil 155–6, **157**, 225
exotic species introductions 50
eye relief 156, 157, 225, 226, 227
eyepieces **153–4**, 154–6, **156**–7, 157, 159, **163**, 163, 165–6, 225, 226, 227
eyes 64, 100, 112, **133**, 140, **147**, 151, **151**, 152, 213
　see also opsin; pigments, retinal
　cone cells 132–3, 135, 144–5, **145**, 146, 214, 225, 228
　cornea **144**, 144, 147–8, 221, 223
　fovea 20–1, **144**, 144, **145**, 145, 225
　lens **144**, 144, 147–8
　Müller cells **145**, 145–6
　retina 132, **144**, 144, 145
　rod cells 4, 132, **133**, 135, 144, 146, 214, 228
　size 144

Falco peregrinus 20
Falco tinnunculus 111
Faraday, Michael 32
Feynman, Richard 35, 41, 58, 204
fields 32, 58–60, 225
　electric 32, 102–3, 110, **114**, 114, **134**, 134, 176, 224, 227, 229
　electromagnetic 33, 40, 100, 102, 222, 227, 229

geomagnetic 31, 61
gluon 58
gravitational 33
　magnetic 32, 61, 62–3, 227, 229
　quantum 58, 225, 228, 231
field of view 143, 155–6, **156**, 157, 162, 165, 225, 225, 227
field stop 155, 157
filtering 78, 80
filters 80, 103, 226
flavin adenine dinucleotide 61, 62, 226
flight, principles of 183–90 *see also* drag
　airflow **184**, 184–6, **186**, **187**, 189–90
　angle of attack 189–90, **190**–1, 191, 221, 231
　buoyancy 183, 222
　downwash 185, 192, 224
　lift 26, 184, 185–6, **186**, 187–90, **190**, 191–3, **193**, 195, 218, 219, 221, 224, 227, 231
　stall **190**, 190–1, 221, 231
　thrust 180, 191, 192–3, 194, 232
　upwash 187, 188, 232
fluid dynamics 183–5, 192
　turbulence 183, 190–1, **191**, 232
　viscosity 189, 233
　vortices 187, 187–8, 198, 199, 233
focal length 139, 140–2, 144, 147, 148, 151, 152, 153
focus 137–8, **139–40**, 140, 141–2, **147**, 147–8, 151, 151, 152, 163, 165–7, 221, 226, 231
forces 67, 225
　electromagnetism 56, 58, 88, 202, 231
　electrostatic 27–30, 45, 75, 85
　geomagnetism 31–2
　gravitation 14, 24–6, 29, 68, 183, 226
　magnetism 30–2, 56, 60
　strong nuclear 56, 58, 99, 202, 231
　weak nuclear 99, 202, 210, 231
Fourier analysis 91, 92, 97, 225
Fourier series 225, 231
Fourier transform 117–18, **118–19**, 119, **120**, 231, 232
　discrete 94–7, 224, 225
　fast 97, 117, 225
　short-time 96–7, 225, 226, 230, 231
fractals 19–21, 39, 47, 203–4, 226
frequency domain 92, 226, 224, 232

Galileo 4, 7, 12, 152
Gallus domesticus 11
gamma rays 100, 101

gannets 67, 68, 179
gauge symmetries 202
Gell-Mann, Murray 55, 57, 206
geometry 21, 118, 120, 127, 140, 155, **178**, 193
  thin film 120, **121–22**, 122–3, **123**, 124
glass 107, 108, 111, 120, 138, 139, 161, 162, 163, 164, 224 *see also* prisms
Gloger's rule 179, 226
Golden Eagle 49
gradients 226
  concentration 43–4, 46, 51–2, 175, 223
  pressure 73–4, 77, 181, 185–77, 190–1, 199
  temperature 181
graphs 12–13, **13**
gravity *see* forces, gravitation
Great Gray Shrike 111
gulls 179
Gymnogyps californianus 57

Hann window 96, **97**, 226, 230
harmonics 76, 113, 183, 225, 226
hawks
  Broad-winged Hawk **144**
  Harris's Hawk 49
  Rough-legged Hawk 111
hearing 81–2, 96, 208
heat 101, 130, 170–3, *see also* energy, thermal; thermoregulation
helium 99
herons 109–10, 111, 191
  Reef Heron 109
Hooded Merganser 147–8, **148**
Horned Guan 87
hummingbirds 24, 27, 29, 124, 161, 195–6, **196**, 202
  Anna's Hummingbird 29, 124
  Calliope Hummingbird 24
hydrogen 56, 99, **126**, 126–7, **127**
hydrophilic materials **177**, 177, **178**
hydrophobic materials **177**, 177, **178**, 217, 226
hyperopia *see* vision, farsightedness

images
  real 150, 153, 226
  virtual 150–1, **151**, **154–5**, 155, 226
inclined planes 67, 160
Indian Roller 119
induction 32
inertia 24, 26, 226, 228

insulation 172, 175–6
insulators 30, 175–6, 224
intensity 75, 83, 100, 117, 136, 208, 227
interactions (electromagnetic, weak and strong)
interference
  constructive 71, **72**, 72, 115, 121, 119, 122, **123**, 161, 209, 227
  destructive **72**, 72–3, 87, 88, 91, 114, 115, 121, 160, 162, 209, 227
inverse square laws 59, 75, 100, 223, 227, 227
ions 42–3, 45–6, 83, 99, 132, 227

keratin 116–17, **117**, 124, 179, 227
kingfishers 66–9

Lanius excubitor 111
least time, principle of 21, 105, 108–9, 211
lenses 139, **139–40**, 140–1, 149–50, **150–1**, 151–2, **152–3**, 153, 227 *see also* eyes
  aspherical 165
  compound 148, 163–5, 223
  negative **152**, 152–3, **155**, 156, 163, 226
  objective **153**, **154**, 154–6, **156–7**, 157, 159, **163**, 165–6, 225
  positive **140**, **150–1**, 152–3, **153–4**, 156, **157**, 163, 226
  spherical 139, **139**, **165**, 165, 231
leverage, mechanical 67, 69, 196
Lévy flights **48**, 48–9, 204
light 40–1, 100, 128, 227
  daylight 13, 37, 64
  dispersion 115, 162–5, 224
  frequency 113–15, 126, 128, 208
  infrared 101, 126, 131, 181, 226
  polarization 64, 102–4, 107, 108, **110**, 110–11, **114**, 229
  polarized 102–3, **103**, 104, **110**, 110–11
  reflectance 161, 164, 215
  reflection 104–7, 106, 107, 108, 110, 115, 116, 120–4, 151, **158**, 158–9, 161–2, **162**, 230
    diffuse 106, **107**, 116
    inverted 121–2, **158**, 158, **160**, **162**, 162
    specular 106, **107**
    total internal 158, **160**, 160, 229, 232
  refraction 108–9, **109–10**, 115, 120–1, 124, 138, **139**, 140, 143, **147**, 147–8, **152**, **163**, 198, 223, 227, 230 *see also* refractive index

scattering 102–3, **103**, 105, **107**, 110, 113, 114–15, 119–20, **120**, 125, 131, 145
    coherent 119, **120**
    incoherent 116, 119, **120**
    Mie 114, 115, 228
    Rayleigh 114, 115, 119, 162, 228, 230
    speed of 107–8, 230
    sunlight 98, 100–2
    ultraviolet 101, 111, 113, 122–3, 125–6, **126**, 128, 129, 179, 180, 181, 211, 229
    unpolarized 103, **103**
    visible 41, 100, **101**, 113, 119, 125, 128, 129, **130**, 162, 208, 223, 227, 228, 229, 230
    wavelength 41, 112–13, 114, 122–3, **123**, 125, **126**, 128–9, **129**, 132, 162–3, 227
local rules 53, 228 *see also* birds, behavior of: flocking
logarithmic spirals **20**, 20–1
Lophodytes cucullatus 147–8, **148**
loudness 75, 227 *see also* sound

Macrocephalon maleo 17
magnets 227 *see also* fields; forces, magnetism
    earth, the 31
    electrical 31
    electro 31, 88, 227
    permanent 31
    magnetic polarity 31
magnetic resonance imaging 61, 207
magnification 141, 152–4, 155, 156, **157**, 157, 227
magnitude 25, 66, 71, 177, 184, 188, 221, 224, 230, 231, 232
Maleo 17
Mallard 197
Mandelbrot set **19**, 19–20, 39
Mandelbrot, Benoit 19, 48
mass 21–2, 24, 25, 26–7, 28, 56–7, 113, 128, 183, 228
mathematics 10–12, 15, 18–20, 22, 23, 28, 91, 183, 202 *see also* counting; numbers
matter
    absorption of light by 105, 125–30, **130**, 131, 221
    states of 66, 170–1
    wave-like character of 58–60
Melanerpes formicivorus 68
Mendeleev, Dmitri 16–17, 55
metabolic processes 171, 172, 180
micro-electromechanical systems 86, 228
microorganisms 40

microphones 85
    cardioid 87, 222
    condenser 85–6, **86**, 223, 228
    directional 86–7
    shotgun 87–8, **88**
microscopy 7, 35
    atomic force 41, 204
    electron 41, 117, 119
microstructures 107, 116, **117–20**, 178, 217, 228
    orderliness of 116–20
migration 13–14, 31–2, 37, 47, 64, 181, 198
Milvus migrans **145**
mitochondria 42, 44
molecules 35, 36, 39, 40–6, 51, 60, 126–30, 170–1, 172, 204 *see also* matter
    isomerization **132**, 132
    isomers 132, 132, 227
    organic 125
    relative size 40–1
    response to light 101–3, 110, 113–15, 128
momentum 22, 24–5, 42, 77, 128, 185, 228
    conservation 23, 25
motion 14, 16, 22, 29, 31, 46, 47, 53, 60, 66, 70–1, 74, 77–8, 82, 85, 103, 113, 131, 162, 170, 171, 183–5, 187, 193–5, 198 *see also* Brownian motion
    equations of 27
    laws of 24–6
murmurations 53, 198, 222, 228
musical instruments 77

Navier–Stokes equation 183, 228
Newton, Isaac 25, 26, 27, 33, 202, 210
nightjars 146
nitrogen 102, 113, 126, 165
Noether, Emmy 22
Noether's theorem 22, 23, 228
nuclear fusion 99–100
numbers 10–11, 12, 38–9, 210 *see also* counting; mathematics
Nyquist's theorem 90, 228

opsin 132, **133**, 228
optical instruments 1–2, 6, 108, 161, 165–7, 228
    see also binoculars; cameras; glass; lenses; telescopes 7, 35, 40, 149, 152, 164, 224
orbitals 57, 59, 60–2, 102, 206, 228
    atomic 125–6
    molecular 126–7, **127**, 223

orders of magnitude 37, 38, 56, 180, 224
Oreophasis derbianus 87
organelles 34–5, 40–1, 124, 228
oscillation 61, 62, 90, 91, 108, 110, **110**, 113, 226, 228, 230
Osprey 48
owls 84, 146
    Northern Saw-whet Owl 84, **84**

Pandion haliaetus 48
parabolas **14**, 14, 27, 87, 165
parabolic dishes 87, 165
Parabuteo unicinctus 49
parrots 23, 127, 129
Passerculus sandwichensis 104
Pauli exclusion principle 59, 228
Pearson, Karl 46
pelicans 111, 179
penguins 171
    Emperor Penguin 146
    Humboldt Penguin 36
    Magellanic Penguin 49
Peregrine Falcon 20
periodicity 16–17, 55, 118
Perissocephalus tricolor 81
Phalaropus tricolor 198–9
pheasants 124
photography *see also* cameras; electronics, light sensors; lenses
    aperture 141, **141**, 142
    ASA ratings 142, 222
    autofocus 166–8, 215, 229
    depth of field 141–2, 223, 224, 226
    exposure 141, 142
    focal length 139, **139–40**, 140–1
    f-stops 142, 226
    ISO ratings 142–3, 213, 222, 227
    photographic film 142, 222
    shutter speed 141, 166
photons 125, 130, 225, 229
physics 3–5, 12, 55
    Newtonian 128, 182, 188 *see also* Newton, Isaac
    quantum mechanics 128, 229
pi 15
pigmentation 52, 125, 128
pigments 125, 127, 129–30, 131, 132, 146, 222, 229
    carotenoids 125, **129**, 129, 222

melanins 124, 125, 129, **130**, 179–80, 223, 226, 228
retinal 132, **132–3**, 228
pixels 117–18, **118**, 135, 143, 213–14
plants, growth and reproduction 13, 27
plumage 21, 51–2, 119, 172, 176, 179–80, 198
Poecile atricapillus 94
polarizability 224, 229, 230
pollen 35
pollination 27, 29–30
potoos 146
power 75, **192**, 192, 193, 195, 196, 229
precession 60–1, 62, 63–4, 229, 231
pressure 42, 100, 178, 199, 222, 223, 229, 232
    see also air; gradients
prisms 159–161, 210
    Abbe–König prism 159, **160**, 221, 230
    Amici 159, **160**
    coatings 160–2, 165, 221, 229, 230
    Porro 159, 229
    roof 159, **160**
    Schmidt–Pechan 160, **161**, 221, 230
Procnias averano 81
Psittacus erithacus 11
Psophodes olivaceus 81

radiation 71, 108, 112, **114**, 124, 126, 128–9, 131, 162, 179, 181, 229 *see also* light; radio frequencies; sound
    electromagnetic 59, 61, 100–1, 112, 125, 226, 228
    solar 101, 110, 180
    thermal 172–3, 175, 181, 226
radio frequencies 60–1, 64
rainbows 115, 162
Ramphastos toco 172–3
random walks 46–7, **47**, 48, 53, 100
raptors 143–6, 184
rarefaction 74, **75**, 82, 230
rate of change 25, 30, 221, 224, 225, 229, 232
ray-tracing diagrams **150**, 150, **151–3**, 153, **154–8, 161**
reaction–diffusion equations 51–2, 53
reaction force 26, 67, 121, 232
Red-legged Partridge 21
refractive indices 108, 120, 124, 138, 139, 147, 161–4, 215, 223, 230
resistance 24, 228
    air 29, 66, 181, 191–2, 218

electrical 230
thermal 180
resolution 20, 38, 83, 90, 95, 142–6, 222, 230
resonance 61, 127, 230
rete mirabile 175, 230

scalar fields 32
scalar quantities 226, 229, 233
scalars 25, 230
scales 38–41, 56
Scolopax rusticola 117
seabirds 42–3
seasons 3, 13, 15, 16
Selasphorus calliope 24
semiconductors 41, 133–4, **134–5**, 229, 230
Shearwater
 Cory's 49
 Scopoli's 49
Sialis sialis 50
signaling states 63–4
signals
 analog 88–9, 89, 92–4, 221–2
 digital 89, **89**, 90–1
silicon 133–4
sine 93, 109, 218, 231
sine waves 14–15, **15**, 16, 40, 70, **73**, 73, 75, 91–2, **92–3**, 93, **96**, 97, 208, 225, 225, 227, 231
 see also waves
sinusoids *see* sine waves
Snell's law 109, 152, 158, 231, 232
snow geese 179
sodium–potassium pump 44–5, **45**
solar angle 13–14, **14–15**, 15, 115
sound 75–8 *see also* audio; intensity; loudness; power
 decibels 75, 208, 224
 recording technology 84–90, 97, 209, 222
 timbre 75–6, 77–8, 82, 85, 86, 94, 226, 232
 waves 74–5
 within a confined space 78–9
sparrows
 Saltmarsh Sparrow 173
 Savannah Sparrow 104
species counts 12–13, **13**
spectra 210, 231
 colored 162, 224
 light 100–1, **101**, 102, 112, 118, 125, **129**, 129, **130**, 136, 162, 210, 213, 224, 228, 229, 230

sound 93–4, **94–5**, 95–6, **96–7**, 97
spectral leakage **96**, 96, 226, 231
spectral lines 126
spectrograms **79**, 80, 91, 95, **95**, 96–7, 117, 224, 231
speed 21, 22, 25, 32, 66, 67–8, 74, 99, 106, 107–9, **109**, 115, 122, 170, 171, 184, 185–6, 189, 192, 195–6, 222, 230, 231 *see also* light, speed of
Spheniscus
 humboldti 36
 magellanicus 49
spherical aberration 165, 231
Spilornis cheela panayensis **53**
Standard Model, the 58–9, 231
starlings 124, 181
 European Starling 50–1, 53, 57
Sterna
 hirundo 193
 paradisaea 37
stress 69, 179
Sturnus vulgaris 50–1, 53
subatomic particles 55, 57–8, 229, 231 *see also* atoms; fields
 bosons 99
 charged 30, 99, 100, 103, 105, 231
 deuterons 99, 210
 electrons 17, 56, 59, 60–1, 101–2, 108, 113–15, 125–7, 131, 134, 206, 225
 valence 126–7, 133, 229, 232
 gluons 228, 229 *see also* fields, gluon
 hadrons 55, 226, 228, 229
 leptons 57
 neutrons 56, 99, 206, 226, 228, 229
 protons 56, 99, 206, 226, 229
 quarks 55–7, 60, 99, 226, 228, 229
 spin 60–2, 225, 231
sun, the 99–100, 224 *see also* energy, solar; radiation, solar; solar angle
 photosphere 100, 101, 103
sunbirds 124
superposition 73, 91, 105, 114, 115, 116, 127, 162, 227, 231
symmetry 16, 23, 231

Taeniopygia guttata 104
tangent 156, 231
telescopes 152, 153–7, 166
 astronomical 87, 157, 165–6

Galilean **154–6**, 157, 226
Keplerian 156, **157**, 157, 159, **163**, 221, 227
temperature 13, 32, 58, 99, 100, 114, 131, 169–71, **171**, 232
tension 77, 80, 232
terns 111, 179
    Common Tern 193
Thalassarche melanophris 49
thermal budget 172, 180
thermodynamics 170, 172, 181
thermoregulation 170–5, 230, 232
thrushes 76
thrust 180, 191, 192, 193, 194, 232
time domain 91, **94–5**, 230, 232
Toco Toucan 172–3
transducers 82, 86–8, 100, 130, 135–6, 223, 232
tribocharging 29
Trochilidae 24
trogons 124
tryptophan 61, 62, 232
turacos 125
Turing patterns 52–3, **52–3**
Turing, Alan 51

vacuums 108, 175
vector fields 32, 64
vector quantities 25, 225, 228, 232
vectors 232
velocity 22, 24–5, 32, 58, 66–7, 70–1, 76, 78, 108, 184, 185–7, 188–9, 190–2, **192**, 193, 195, 196, 199, 207, 232
venae comitantes 175, 233
vignetting 155, 157
vision 38, 112, 131–2, 144, 151–2
    accommodation 148, 221, 226
    acuity 20, 143–4, 146, 199, 225
    far sightedness 149, 151, **151**, 226
    nearsightedness 151
von Humboldt, Alexander 35–6

water 35, 40, 42, 114, 176–7
    penetration 178–9

repellency 176–9, 217
waves 70, 208
wavenumber 117, 233
waves 69–70, **70**, 71–3, 128, 233 *see also* interference; sine waves; superposition
    amplitude 16, **70**, 70–1, **72–3**, 73, 75, 78, 79, 83, 89, **89**, 90, 91, 92, **92**, **94**, 95, **95**, 113, 127–8, 221, 226, 230, 231, 233
    frequency 61, 70, 73, 76, 78–80, 83, 87, 88, 90, 91–3, **94–5**, 95–6, 101, 102, 113, 115, 128, 162, 224, 225, 226, 230, 230, 233
    longitudinal 74
    microwaves 101, 226
    non-trivial 92, **92**
    periodic 16, 70–1, 90, 92, **92**
    propagation direction 86, 102, 110, 224, 229, 230
    transverse 74, 229
    wavelength 16, 40–1, 59, **70**, 70–1, **72–3**, 75, **78**, 78, 87, **88**, 88, **93–3**, 93, 95, 100–2, 114, 116, 122–3, 125–6, 127–9, **130**, 132, 136, 162–3, 165, 206, 209, 214, 222, 226, 227, 228, 230, 233
weight 24, 26, 183, 188–9, **189**, **193**, 193, 222, 227, 233
Wilson's Phalarope 198–9
winglet 187, 233
wings 231
    loading 188–9, **189**, 233
    shape **184**, 184–5, 186, 187–8
woodpeckers 68–9
    Acorn woodpecker 68
    Ivory-billed woodpecker 69, 85
work 22, 67, 85, 225, 229, 233
Wrybill 23

X-rays 101

Zebra Finch 104